FEYNMAN'S
LOST
LECTURE

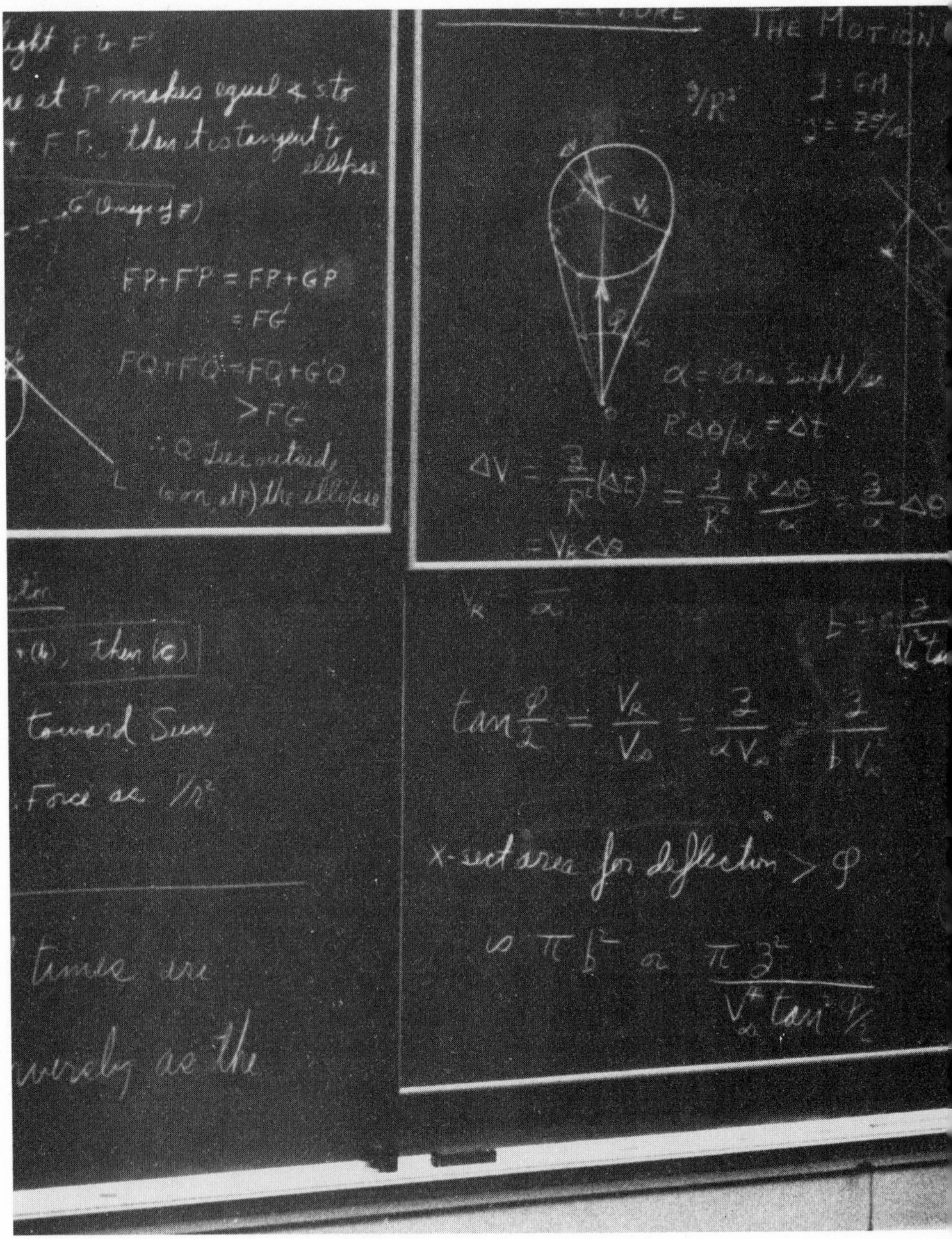

...ight F to F'

...e at P makes equal ∠'s to

... F T, then t is tangent to
ellipse

___ (image of F)

$$FP + F'P = FP + GP$$
$$= FG$$
$$FQ + F'Q = FQ + G'Q$$
$$> FG$$

∴ Q lies outside,
(or on it) the ellipse

...tion

...(b), then (c)

toward Sun

Force α $1/r^2$

...times are

...ersely as the

---

g/R^2 $J = GM$
$g = 75 \%$

$$\alpha = \text{Area swept/sec}$$
$$R \Delta\theta / \alpha = \Delta t$$

$$\Delta V = \frac{J}{R^2}(\Delta t)^2 = \frac{J}{R^2} \frac{R^2 \Delta\theta}{\alpha} = \frac{J}{\alpha} \Delta\theta$$
$$= V_R \Delta\theta$$

$$V_R = \frac{J}{\alpha}$$

$$\tan\frac{\varphi}{2} = \frac{V_R}{V_0} = \frac{J}{\alpha V_0} = \frac{J}{b V_\alpha}$$

x-sect area for deflection $> \varphi$

$$\alpha \pi b^2 \quad \alpha \quad \pi \frac{J^2}{V_0^4 \tan^2 \frac{\varphi}{2}}$$

FEYNMAN'S LOST LECTURE

The Motion of Planets Around the Sun

DAVID L. GOODSTEIN AND JUDITH R. GOODSTEIN

W · W · NORTON & COMPANY

NEW YORK   LONDON

Copyright © 1996 by the California Institute of Technology

Printed in the United States of America
First published as a Norton paperback 1999

For information about permission to reproduce selections from this book,
write to Permissions, W. W. Norton & Company, Inc., 500 Fifth Avenue,
New York, NY 10110.

The text of this book is composed in Times Roman,
with the display set in Times Roman Bold and Caslon Openface.
Composition by Crane Typesetting Service, Inc.
Manufacturing by The Courier Companies, Inc.
Book design by Margaret M. Wagner

Library of Congress Cataloging-in-Publication Data
Goodstein, David L., 1939–
Feynman's lost lecture : the motion of planets around the sun / by David L. Goodstein
and Judith R. Goodstein.
p.   cm.
Includes bibliographical references and index.
ISBN 0-393-03918-8
1. Planets—Orbits.   2. Celestial mechanics.   3. Conic sections.
4. Feynman, Richard Phillips.   I. Goodstein, Judith R.   II. Title.
QB603.M6G66   1996
521′.3—dc20   95-38719

ISBN 0-393-31995-4 pbk.

W. W. Norton & Company, Inc, 500 Fifth Avenue, New York, N.Y. 10110
www.wwnorton.com

W. W. Norton & Company Ltd., 10 Coptic Street, London WC1A 1PU

1 2 3 4 5 6 7 8 9 0

To the memory of R.P.F.,
who would be appalled that we found it
necessary to explain what he had
said so clearly.

Contents

Preface

This is the story of how Feynman's lost lecture came to be lost, and how it came to be found again. In April 1992, as Caltech's archivist, I was asked by Gerry Neugebauer, the chairman of the Physics Department, to go through the files in Robert Leighton's office. Leighton was ill and had not used his office for several years. Marge Leighton, his wife, had told Neugebauer that it was all right to clean out the office— she'd already collected her husband's books and personal effects. I could take what I wanted for the archives, and the Physics Department would dispose of the rest.

Besides heading the Physics Department from 1970 to 1975, Leighton, together with Matthew Sands, had overseen the editing and publication of Richard Feynman's two-year course of lectures in introductory physics, delivered to Caltech freshmen and sophomores. The lectures, published in the early 1960s in three volumes by Addison-Wesley, dealt with virtually every subject in physics, with a point of view that remains fresh and original to this day. I was hoping to find some tangible evidence of the Leighton-Feynman collaboration.

It took me a couple of weeks to sift through the stacks of paper, which were stashed everywhere, but Leighton didn't disappoint me. I unearthed two folders, one marked "Feynman Freshman Lectures, unfinished,'' another labeled ''Addison-Wesley,'' wedged between budget sheets and purchase orders from earlier decades and reams of yellowing computer paper covered with endless columns of numbers, all thrown together in a storage closet just outside his office. Leighton's correspondence with the publisher contained details about the format, the color of the cover, comments by outside readers, adoptions at other schools, and estimates of how well the volumes would sell. That folder I put in the ''Save'' pile. The other folder, the one containing the unedited Feynman physics lectures, I carried back to the archives myself.

In his June 1963 preface to *The Feynman Lectures on Physics*, Feynman commented on some of the lectures not included there. He'd given three optional lectures in the first year on how to solve problems. And, indeed, three of the items in Leighton's folder turned out to be the raw transcripts for Reviews A, B, and C, offered by Feynman in December 1961. A lecture on inertial guidance, which Feynman gave the following month, didn't make the cut either—an unfortunate decision, according to Feynman—and I found a partial transcript of this lecture in Leighton's folder. The folder also contained the unedited partial transcript of a later lecture, dated March 13, 1964, along with a sheaf of notes in Feynman's handwriting. Entitled ''The Motion of Planets Around the Sun,'' it was an unorthodox approach to Isaac Newton's geometric demonstration of the law of ellipses, in the *Principia Mathematica*.

In September 1993, I had occasion to draw up a list of the original audiotapes of the Feynman lectures, which had also been contributed to the archives. They included five lectures that were not to be found in the Addison-Wesley books. Then I remembered the five unpublished lectures in Leighton's file; sure enough, the unedited transcripts matched the tapes. The archives also had photographs of the blackboard diagrams and equations for four of these lectures—the four mentioned by Feynman in his preface—but I could find none for the March 1964 lecture on planetary motion. (In the course of selecting illustrations for this book, I did stumble upon one photograph of Feynman taken during this special lecture. It is reproduced here as the frontispiece.) Although Feynman had given Leighton his notes on the 1964 lecture, which included sketches of his blackboard drawings, Leighton apparently decided not to include it in the last (1965) volume of *The Feynman Lectures on Physics*, which dealt primarily with quantum mechanics. In time, this lecture was forgotten. For all practical purposes, it was lost.

The idea of rescuing all five unpublished Feynman lectures from oblivion appealed to David and me. So the following December, when we went, as we often do, to the Italian hill town of Frascati, we took along copies of the tapes, the transcripts, the blackboard photographs, and Feynman's notes. In the course of the next two weeks, we listened to the tapes, took notes, laughed at the jokes, strained to hear the students' questions and Feynman's answers after each lecture was over, took more notes. But in the end, we decided that the only lecture that still had the vitality, originality, and verve we associated with Feynman's presence in the classroom was the 1964 lecture on planetary motion— the one lecture that demanded a full complement of blackboard photographs. And we didn't have them. Reluctantly, we abandoned the project.

Or so I thought. As it turned out, bits and pieces of the lecture haunted David, especially when he came to teach the same material in freshman physics the following year. He had the tape. But could he reconstruct the blackboard demonstrations from the few tantalizing sketches in Feynman's notes and the few words Feynman had jotted down more for himself than for the students? "Let's try again," he announced, early in December 1994, as we were packing for a trip through the Panama Canal. This time, we would take along only the transcript of the 1964 lecture, the lecture notes, and selected pages from Kepler's *The New Astronomy* and Newton's *Principia* for good measure.

It took the S.S. *Rotterdam* eleven days to sail from Acapulco to Fort Lauderdale. For two to three hours each day, David would hole up in our cabin and work on deciphering Feynman's lost lecture. He began, as Feynman had, with Newton's geometrical proofs. The initial break came when he was able to match up Feynman's first sketch with one of Newton's diagrams, on page 40 of the Cajori edition of the *Principia*. We'd been at sea for three, maybe four days, Costa Rica's shoreline plainly visible, when David announced that he, too, could follow Newton's line of reasoning up to a point. By the time we'd exchanged the Pacific Ocean for the Atlantic, he was completely absorbed in Feynman's sparse, neatly labeled pencil drawings of curves and angles and intersecting lines. He stayed in the cabin, ignoring the scenery in favor of geometric figures—Newton's, Feynman's, and his own—longer and longer each morning and in the evening as well. When we arrived in Fort Lauderdale, on December 21, he knew and understood Feynman's entire argument. On the plane home, the book took shape.

Its final form owes much to the contributions of family and friends. Marcia Goodstein ingeniously outwitted simple-minded software to produce the nearly 150 figures needed to tell Feynman's geometric tale. Sara Lippincott, skilled editor and diplomat, gently massaged the prose and the presentation. Ed Barber, vice-chairman at W. W. Norton, invested years of friendly persuasion, which paid off when the lost lecture showed up. Robbie Vogt supplied the story of how it originated. Jim Blinn read the manuscript and made helpful suggestions. Valentine Telegdi called our attention to the proof by James Clerk Maxwell. Finally, we wish to thank Mike Keller, Caltech's intellectual property lawyer, for his cheerful help. Proceeds from this book will be used to support scientific and scholarly research at Caltech.

All the photographs in this book are taken from the Caltech archives.

J.R.G.
Pasadena, May 1995

FEYNMAN'S
LOST
LECTURE

Introduction

I would rather discover a single fact, even a small one, than debate the great issues at length without discovering anything at all.

—GALILEO GALILEI

This is a book about a single fact, although certainly not a small one. When a planet, or a comet, or any other body arcs through space under the influence of gravity, it traces out one of a very special set of mathematical curves—either a circle or an ellipse or a parabola or a hyperbola. These curves are known collectively as the conic sections. Why in the world does nature choose to trace out in the sky those, and only those, elegant geometrical constructions? The problem turns out to be not only of profound scientific and philosophical significance but of immense historical importance as well.

In August of 1684, Edmund Halley (after whom the comet would be named) journeyed to Cambridge to speak to the celebrated but somewhat strange mathematician Isaac Newton about celestial mechanics. The idea was abroad in scientific circles that the motions of the planets might be a consequence of a force from the Sun that diminished as the inverse square of the distance between the Sun and the planets, but no one had yet been able to produce a satisfactory demonstration. Yes, Newton let on, he had been able to demonstrate that such a force would give rise to elliptical orbits—exactly what Johannes Kepler had deduced some seventy years earlier from observations of the heavens. Halley urged Newton to let him see the demonstration. Newton apparently begged off, saying he had misplaced it, but promised to work it out again and send it to Halley. In fact, a few months later, in November 1684, Newton did send Halley a nine-page treatise in which he demonstrated that an inverse-square law of gravity, together with some basic principles of dynamics, would account for not only elliptical orbits but Kepler's other laws of planetary motion as well, and more besides. Halley knew that he held in his hands nothing less than the key to understanding the universe as it was then conceived.

He urged Newton to let him arrange for its publication. But Newton was not entirely satisfied with his work and delayed, wanting to make revisions. The delay lasted almost three years, during which Newton, now thoroughly hooked on the problem, seems to have done nothing else but work on it. What emerged at the end, in 1687, was *Philosophiae Naturalis Principia Mathematica*, Newton's masterpiece and the book that created modern science.

Nearly three hundred years later, the physicist Richard Feynman, apparently for his own amusement, undertook to prove Kepler's law of ellipses himself, using no mathematics more advanced than elementary plane geometry. When he was asked to give a guest lecture to the Caltech freshman class in March 1964, he decided to base it on that geometric proof. The lecture was duly recorded on audiotape and transcribed. Usually, photographs were made of the blackboard during Feynman's lectures; however, if they were made in this case, they have not survived. Without any indication of what geometric diagrams he was referring to, the lecture was incomprehensible. But when Feynman's own notes for that lecture were rediscovered among the papers of his colleague Robert Leighton, it became possible to reconstruct his entire argument.

The discovery of Feynman's lost lecture notes affords us an extraordinary opportunity. For most people, Feynman's fame rests on his picaresque exploits, recounted in two anecdotal books (*"Surely You're Joking, Mr. Feynman!"* and *"What Do You Care What Other People Think?"*) which he produced late in life in collaboration with Leighton's son, Ralph. The stories in these books are amusing enough, but they take on a special resonance because the protagonist was also a theoretical physicist of historic proportions. Yet for the nonscientist reader there is no way to peer into Feynman's mind and see that other side of him—the powerful intellect that left an indelible imprint on scientific thought. In this lecture, however, Feynman uses all his ingenuity, insight, and intuition, and his argument is not obscured by the layers of mathematical sophistication that made most of his accomplishments in physics impenetrable to the uninitiated. This lecture is an opportunity for anyone who has mastered plane geometry to see the great Feynman at work!

Why did Feynman undertake to prove Kepler's law of ellipses using only plane geometry? The job is more easily done using the powerful techniques of more advanced mathematics. Feynman was evidently intrigued by the fact that Isaac Newton, who had invented some of those more advanced techniques himself, nevertheless presented his own proof of Kepler's law in the *Principia* using only plane geometry. Feynman tried to follow Newton's proof, but he couldn't get past a certain point, because Newton made use of arcane properties of conic sections (a hot topic in Newton's time) that Feynman didn't know. So, as he says in his lecture, Feynman cooked up a proof of his own.

Moreover, this is not just an interesting intellectual puzzle that Feynman has doodled with. Newton's demonstration of the law of ellipses is a watershed that separates the ancient world from the modern world—the culmination of the Scientific Revolution. It is one of the crowning achievements of the human mind, comparable to Beethoven's symphonies, or Shakespeare's plays, or Michelangelo's Sistine Chapel. Aside from its immense importance in the history of physics, it is a conclusive demonstration of the astonishing fact that has mystified and intrigued all deep thinkers since Newton's time: nature obeys mathematics.

For all these reasons, it seems worthwhile to open Feynman's lecture for the world to see. The price, for the reader, is not cheap. This particular lecture must have been daunting even for the mathematical whizzes in Caltech's freshman class. Even though each individual step is elementary, the proof taken as a whole is not simple. And at one remove from Feynman's blackboard and his vivid presence in the classroom, the lecture becomes a good deal more difficult to follow. Nevertheless, it is the purpose of this book to draw the reader in by describing the historical significance of Newton's demonstration of the law of ellipses, and of Feynman's own life and work; and then to do no less than reconstruct the proof that Feynman set out in his lecture, and explain it so meticulously that readers who remember what they were taught in high school geometry will understand Feynman's brilliant formulation. Then the reader will be ready for the written and audio versions of the lecture itself, which are included with this book.

1

From Copernicus to Newton

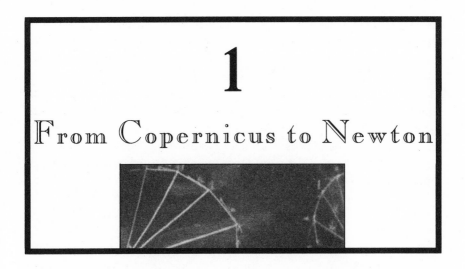

In 1543, as he lay on his deathbed, the Polish cleric Nicolaus Copernicus was shown the first copies of his book, *On the Revolutions of the Celestial Spheres*. He had purposely delayed its publication until he no longer would have to face the consequences. The book proposed the unthinkable: that the Sun, not the Earth, was the center of the universe. The book was about revolutions, real ones in the sky, and it kicked off what came to be called, metaphorically, the Scientific Revolution. Today, when we refer to political and other upheavals as revolutions, we are paying homage to Copernicus, whose book about revolutions started the first revolution.

Before Copernicus, our view of the world was derived from the ancient Greek philosophers and mathematicians, frozen in time in the teachings of Plato and Aristotle, who lived and taught in the fourth century B.C. In the Aristotelian world, all matter was made of four elements: earth, water, air, and fire. Each element had its natural place— earth, surrounded by water, at the center of the universe, then air and fire, in ascending spheres. Natural motion consisted of elements seeking their natural places. Thus, heavy, predominantly earthy bodies tended to fall, while bubbles would rise through water and smoke would rise through air. All other motions were violent, requiring a proximate cause.

For example, an oxcart would not move unless an ox was pulling it. Outside the spheres of earth, water, air, and fire, the heavenly bodies revolved on crystal spheres of their own. The heavenly spheres, where only the perfection of circular motion was permitted, were serene, harmonious, and eternal. Only down here on Earth was there change, death, and decay. It was a coherent system of the world, unmistakably designed to put us in our place, but that place was the center of the universe, and we could, for all our faults, easily imagine ourselves to be the purpose of creation. "We were quite happy with Aristotle's cosmos," remarks a character in Tom Stoppard's *Arcadia*, a play that pokes fun at historians and scientists alike. "Personally I preferred it. Fifty-five crystal spheres geared to God's crankshaft is my idea of a satisfying universe."

But there were a few problems even in the serene heavens of the Aristotelian cosmos. The Sun, the Moon, and the stars executed their motions faithfully enough (for the most part), but a small number of prominent bodies called planets (from the Greek for "wanderer") failed to behave themselves properly. Predicting the positions of these bodies—where they would appear in the sky on any given night—was the professional responsibility of astronomers. The information was of some importance for agriculture, for navigation, and, above all, for casting horoscopes in a world deeply steeped in astrology. The idea that the planets went around the Earth in perfect circles didn't accord with observation, but Plato had said that in the heavens only circular motion was possible. So astronomers concocted the scheme of having the planets move in circles, called epicycles, that were themselves centered on other circles, called deferents. If an observation of a planet in the sky didn't quite fit the existing system of deferents and epicycles, another epicycle could be added to refine the calculations and improve the accuracy of the predictions—a practice known as "saving the appearances." This ancient system of astronomy was codified in the second century A.D. by the Alexandrian Greek astronomer Ptolemy, in a book called the *Almagest*. The *Almagest* remained the principal textbook of astronomy for fourteen hundred years, until the time of Copernicus.

In his book, Copernicus pointed out that the whole balky system of epicycles and deferents could be simplified somewhat if, strictly as a mathematical convenience, one placed the Sun rather than the Earth at the center of things. That was the first chapter. The rest of the book is filled with astronomical tables, calculated using epicycles and deferents centered on the Sun. The ruse of mathematical convenience fooled no one—but, on the other hand, very few paid any attention to Copernicus at all in the decades following his death, and even fewer actually bothered to read the book. It is true that during this period Jesuit missionaries were teaching the Copernican system in China, but at its headquarters in Rome the Church was much more concerned with Martin Luther than with Nicolaus Copernicus. Nevertheless, there were a few who noticed, and who cared. Three men in particular were destined to play crucial roles in overthrowing the geocentric universe. They were Tycho Brahe, Johannes Kepler, and Galileo Galilei.

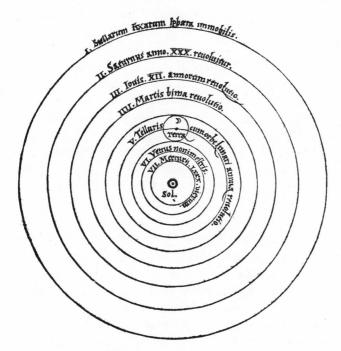

Copernicus's view of the solar system, from *De revolutionibus orbium coelestium*, 1543.

Tycho Brahe (1546–1601) was a Danish nobleman, who as a boy learned to his amazement that it was possible to predict events in the sky, such as the solar eclipse of August 21, 1560, and then learned to his even greater amazement, while observing a conjunction of Jupiter and Saturn in August 1563, that the astronomical tables (including the Copernican tables!) were off by several days—presumably for lack of accurate astronomical data.

After studying law, traveling in Europe, losing his nose in a duel, and replacing it with one of gold, silver, and wax, Tycho scandalized Danish society by marrying a commoner and becoming an astronomer. He set up a small observatory on some family lands, and there, on November 11, 1572, he discovered a brilliant new star where none had been before, in the constellation Cassiopeia. New stars were not supposed to appear in the immutable Aristotelian heavens. His book *De nova stella* (*On the New Star*) offended the Church, established his reputation, and earned him the patronage of Denmark's king, Frederick II.

Frederick gave Tycho the island of Hveen, near Copenhagen, and the financial support to build there the greatest astronomical observatory the world had ever seen. Giant measuring instruments were constructed—Tycho's "great equatorial armilla" was some nine feet across; the diameter of his "great mural quadrant" measured thirteen feet— to make sitings of unprecedented precision, together with magnificent buildings in which to live and work, printing presses to publish new findings, and much more. Tycho called the place Uraniborg, after Urania, the muse of astronomy. Begun in 1576, it operated until 1597. Just a few years later, in 1610, the invention of the telescope would end forever this kind of naked-eye astronomy. Nevertheless, the observations made at Uraniborg during its brief period of existence would decrease the uncertainty of astronomical tables from ten minutes of arc to two minutes of arc. (If you hold your index finger up at arm's length, it covers an angle of about one degree. Ten minutes of arc is one-sixth of that; two minutes of arc is another five times smaller.)

In 1588, Frederick II died, to be succeeded by his son, Christian IV. Christian was irritated by Tycho's incessant demands for lavish support, and by 1597 the situation had deteriorated to the point where Tycho felt obliged to close up Uraniborg, leave Denmark, and establish himself in Prague, where he became imperial mathematician to Rudolph II, king of Hungary and Bohemia and Holy Roman Emperor.

Tycho Brahe at age forty. Frontispiece from Tycho's *Astronomiae instauratae mechanica*, 1602.

By the time Tycho departed for Prague, he had made an indelible contribution to astronomy. That did not satisfy him. The job that still lay before him was to put his precious (and still largely secret) observations to the service of the new cosmology. Not, however, to the cosmology of Copernicus, and certainly not to that of Ptolemy; Tycho had invented a cosmos of his very own. In the Tychonic universe, all the planets revolved around the Sun, and the Sun with the other planets revolved around the Earth, which was restored to the center of the universe. To the modern eye, the Tychonic universe seems to be a compromise between Aristotle and Copernicus, but in his own time Tycho's cosmos was in some ways an even more audacious departure from Aristotle than Copernicus's had been, because it smashed the crystal spheres that were supposed to fill the heavens regardless of whether the Earth or the Sun was at the center. The question was: Would the Tychonic data support the Tychonic universe? To answer the question required a mathematical talent far greater than that possessed by the imperial mathematician. In all of Europe, there may not have been more than one mathematician with the necessary abilities. But at least there was one. His name was Johannes Kepler.

Kepler was born in 1571, the son of a mercenary soldier who quickly evaporated from the scene and a shrewish innkeeper's daughter who would later be tried for witchcraft. Small in stature, fragile in health, and poor, Kepler's obvious keen intelligence nevertheless won him a scholarship that permitted him to attend the University of Tübingen. There he studied under one of Europe's earliest advocates of the Copernican system, Michael Mästlin. After he had obtained his bachelor's and master's degrees, the Tübingen faculty rescued him from a career as a Lutheran minister by recommending him for a post teaching high school mathematics in the Austrian town of Graz.

According to legend, one day in the summer of 1595, Kepler's body was lecturing about geometry to a class of bored adolescents while his mind went rummaging through the tabulated data of Copernican astronomy, his lifelong passion. Inscribing circles inside and outside an equilateral triangle, he suddenly realized that the ratio of the diameters of the two circles (the outer one is just twice as big as the inner) was essentially the same as the ratio of the diameters of the orbits of Jupiter and Saturn. The discovery sent Kepler himself into orbit. He quickly devised a model in which the six invisible spheres that regulated the orbits of the six planets then known were fitted on either side of each of the five "perfect solids" of antiquity (solids having all sides the same: the tetrahedron, cube, octahedron, dodecahedron, and icosahedron), nested one inside the other. Sure enough, by arranging the solids in the right order, the diameters of the spheres came out to be in almost the same ratios as those of the orbits of the planets.

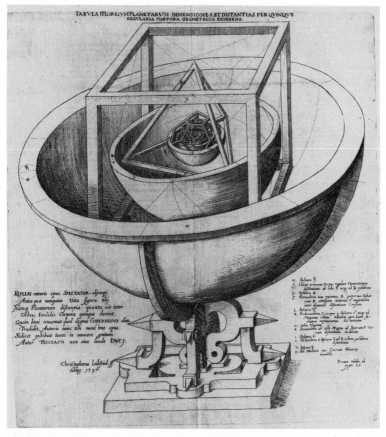

The nested solids (the outermost sphere is Saturn's), from Johannes Kepler's *Mysterium cosmographicum*, 1596.

Kepler's model explained why there were six, and only six, planets—because there were five, and only five, perfect solids—and why their orbits had the ratios they did. The whole arrangement fitted miraculously together. Kepler thought, not for the last time in his life, that he had seen into the mind of the Creator. In 1596, he published his inspiration in the book *Mysterium cosmographicum*, which brought him to the attention of Tycho Brahe.

Tycho was not enchanted by Kepler's Copernican views, but he was impressed with his mathematical talent. He invited Kepler to join him in Prague. Kepler had by now built up a considerable reputation as a skillful astrologer (his predictions of plague, famine, and Turkish invasion often turned out to be correct), but his finances were still precarious, and as a Lutheran he felt persecuted in Catholic Graz. On the first day of January in the year 1600, Johannes Kepler set out to join the Danish astronomer in Prague.[1]

The timid Johannes Kepler did not get on well with the boisterous, metal-nosed Tycho Brahe, but they needed each other. Kepler needed Tycho's data to do his life's work, and Tycho needed Kepler's genius to organize his observations and provide confirmation of the Tychonic universe. The mismatch lasted eighteen months, until, in 1601, Tycho Brahe died suddenly of an acute urinary infection. It is reported that his last words to Kepler were "Let me not seem to have lived in vain." But Kepler, the dedicated Copernican, had no intention of pursuing Tychonic cosmology.

After Tycho's death, Kepler managed, with difficulty (nothing in Kepler's life came easy), to be named his successor as imperial mathematician—a title that turned out to be more honorific than remunerative—and to wrest the superb Tychonic data from Tycho's heirs. He also published a book on astrology. (He regarded all other astrologers to be charlatans and frauds, but for his own part he couldn't suppress entirely the feeling that there might be a certain harmony between human destiny and the heavenly panorama.) And in 1604, while observing a

[1]Why, the reader may ask, do we always refer to Tycho by his first name and Kepler by his last (for example, Tychonic universe, but Kepler's laws)? There is no obvious answer. It may be because Johannes is too common and Brahe is unfamiliar. We also refer to Galileo by his first name, but in his case it hardly matters, because his first name was the same as his last.

rare conjunction of Mars, Jupiter, and Saturn, he saw the appearance of a supernova, a new star that remained visible in the sky for seventeen months.

Kepler's greatest struggle was his "war on Mars"—his attempt to find an orbit for the planet that would be consistent with Tycho's observations. A circle could be fitted to the orbit of Mars if the uncertainty in the observations was ten minutes of arc, as it had been before Tycho. But Tycho's magnificent legacy required something different. Kepler calculated prodigiously, first using an ingenious method to deduce the orbit of the Earth, the uncertain celestial platform from which Tycho's observations had been made. The Earth's orbit could be described well enough by a circle with the Sun slightly displaced from the center. But not the orbit of Mars. Try as he might, no circle would fit. In *Astronomia nova* (*The New Astronomy*), published in 1609, Kepler quotes Virgil to describe his quest:

> Galatea seeks me mischievously, the lusty wench:
> She flees to the willows, but hopes I'll see her first.

In the Copernican system, the Earth is one of the planets. But Earth, being the site of change, death, and decay, is obviously not in a state of Platonic perfection, as planets had been supposed to be—so maybe the orbits of the planets did not need to be Platonic circles at all! ("Oh ridiculous me!" says Kepler, of his failure to grasp this point earlier; we don't write scientific papers that way anymore.) The orbit of Mars was not a circle. It was an ellipse, with the Sun at one focus (a word adopted for the purpose by Kepler from the Latin for "fireplace").

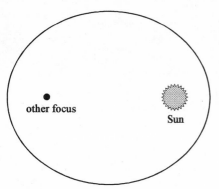

An ellipse with the Sun at one focus (the orbit of Mars is much more nearly circular than this ellipse).

The ellipse is a closed geometric curve known from antiquity. Apollonius of Perga (c. 262–c. 190 B.C.) showed that intersecting a cone and a plane produced two closed curves, a circle and an ellipse,

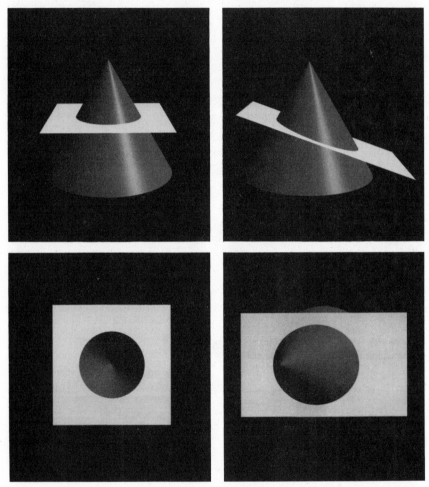

LEFT. plane intersects cone, producing a circle when seen head-on *(below)*. RIGHT. tilted plane intersects cone, producing an ellipse when seen head-on *(below)*.

and two open curves, the parabola and the hyperbola.

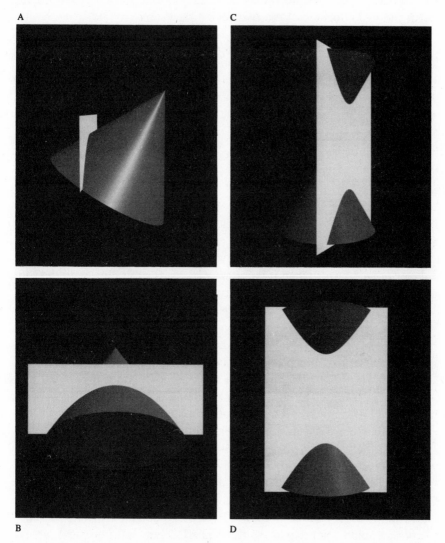

LEFT. plane intersects cone parallel to the opposite side of the cone. Seen head-on the result is a parabola *(below)*. RIGHT. plane intersects both branches of an extended cone. Seen head-on, the result is a hyperbola *(below)*. Unlike the other conic sections, the hyperbola always has two branches.

Collectively, these figures are known as the conic sections. The ellipse, in particular, may be drawn by a string-and-thumbtacks construction, with the tacks located at the two foci:

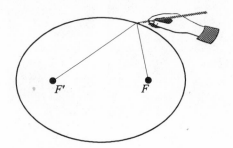

We'll get back to the special properties of the ellipse in Chapter 3.

In *Astronomia nova*, Kepler tells us that in fact the orbits of *all* the planets are ellipses with the Sun at one focus—a statement that became known as Kepler's first law, the law of ellipses. He also tells us that a planet moves faster when it is in that part of its orbit closest to the Sun and slower when it is farther away. Moreover, this speeding up and slowing down of the planet's motion has a most peculiar kind of regularity about it: a line from the Sun to the planet would sweep out equal areas in equal times. This became known as Kepler's second law. Ten years later, in 1619, Kepler published another book, *Harmonices mundi* (*Harmony of the World*), in which he expounded yet a third law. The first two laws describe the motion of a single planet in its own orbit. The third law compares the orbits of the planets. It says that the farther a planet is from the Sun, the slower it moves in its orbit. In particular, a year in the life of a planet (the amount of time it takes to make one complete orbit) is proportional to the three-halves power of the size of the orbit (technically, the longer diameter of the ellipse). Together, these three statements are Kepler's great contribution—Kepler's three laws of planetary motion. In 1627, Kepler published the *Rudolphine Tables*, named for his patron, Rudolph II. These astronomical tables, based on Tycho's meticulous observations combined with Kepler's three laws, made astronomy a hundred times more accurate than it had ever been before.

Frontispiece, *Tabulae Rudolphinae*, 1627. Designed by Kepler, this elaborate engraving depicts the giants of astronomy gathered in the temple of Urania. Kepler put himself and the titles of four of his books in the left panel on the base of the temple.

At about the same time, in Italy, Galileo Galilei wrote in *Il Saggiatore*, "The book of nature lies continuously open before our eyes (I speak of the universe) but it can't be understood without first learning to understand the language and characters in which it is written. It is written in mathematical language, and its characters are geometrical figures." Galileo was not writing to celebrate Kepler's laws, which, ironically, he never even acknowledged, much less embraced. He was, however, writing in defense of the Copernican system. In 1616, the Catholic Church's chief theologian, Cardinal Robert Bellarmine, had declared Copernicanism to be "false and erroneous" and had placed Copernicus's book on the index of forbidden books. Now, however, a new pope, Urban VIII, long a friend and supporter of Galileo's, had been installed, and Galileo hoped to divert the Church from its disastrous collision course with science. He would not succeed.

Galileo was born in Pisa, in 1564, the son of a musician, Vincenzio Galilei. (It was fashionable at the time among Tuscan families to give the firstborn child the family surname as a first name.) Galileo studied medicine at the University of Pisa, but withdrew without graduating, for lack of funds. He taught himself mathematics, published a few essays, and secured a position lecturing on mathematics at Pisa. While at Pisa, he discovered the law of the pendulum (a pendulum takes the same amount of time for a complete cycle, no matter how big or small its arc) and the law of falling bodies (all bodies, regardless of mass, fall with the same constant acceleration in a vacuum); and he did a series of kinematical experiments, using balls and inclined planes, that amounted to nothing less than the invention of experimental science as we know it today. (The title of his book *Il Saggiatore* is usually rendered in English as *The Assayer*, but the modern term "The Experimentalist" more accurately describes what he had in mind.) He apparently embraced Copernicanism early in life, but kept his belief hidden for fear of ridicule. In one of his rare letters to Kepler—really a thank-you note for a copy of Kepler's *Mysterium cosmographicum*—he wrote in 1597, "I indeed congratulate myself on having an associate in the study of Truth who is a friend of Truth." Truth with a capital T is a disguised but unmistakable reference to Copernicus.

The Copernican system, however, was not only an affront to Aristotelian and Church dogma; it seemed also to be an affront to common sense. Any fool could plainly see that the Earth was solidly at rest. If the Earth was spinning on its axis and hurtling through space, as Copernicans claimed, why couldn't we sense all that motion? To put a sharper point on the question, consider the following thought experiment: suppose someone were to drop some heavy object from the top of the Tower of Pisa. Regardless of our cosmological orientation, we can all agree on this much at least: the object will fall straight down to the foot of the tower (ignoring for the moment the tower's famous tilt). But according to the Copernicans the Earth is spinning on its axis as the object falls. If the force of gravity causes the object to fall directly toward the center of the Earth, the object should fall straight down, while the tower rotates away. How far away? It takes an object about two seconds to reach the ground if dropped from the top of the tower. Given the size of the Earth and the fact that it makes a complete rotation once each day, the distance is not hard to compute. The tower should move about half a mile while the object is falling! In other words, if Copernicus was right and the Earth made a complete rotation on its axis each day, then an object dropped from the top of the Leaning Tower of Pisa should hit the ground half a mile away. The fact that it does no such thing seems a pretty decisive refutation of Copernicanism.

The problem for Copernicans in the sixteenth century was not only that such objections were difficult to answer but—even worse—that there seemed to be no starting point for formulating an answer. When Copernicus tore the Earth from the center of the universe, he also tore the heart out of Aristotelian mechanics, the intellectual glue that held everything together. For example, why should a heavy object fall at all, if it was not seeking its natural place? To answer that it fell because of gravity, as was done and still is, merely gives the mystery a name. For believers in Copernicus, the Aristotelian world lay in ruins, and there was nothing to take its place. This was the dilemma that confronted Galileo.

To find out how the world actually works, Galileo conceived the idea of doing experiments whose results could be analyzed using mathematics. It was an idea that would change forever the course of human history. He could not study falling bodies directly, because they fell too fast and there were no good clocks: the first accurate timepieces, based on his own discovery of the isochrony of the pendulum, would come much later. To slow down the motion of falling bodies, he measured the time taken by balls rolling down gently inclined planes—planes that were made as smooth as possible to minimize friction. (Replicas of these instruments, made by skilled craftsmen, may be found in the Museum of the History of Science in Florence.) He tried many schemes to measure accurately the elapsed time while the balls were rolling. The best of these was a kind of water chronometer. Water was allowed to run through a tube—which he could open and close by means of his finger—into a second container while the ball was rolling. Then he would weigh the water that had run out. The weight of the water would be proportional to the elapsed time. Modern reproductions of his experiments have shown that with practice he could achieve in this way a precision of about two-tenths of a second. Better measurements of elapsed time were seldom made until the twentieth century.

Using this technique, Galileo discovered his law of falling bodies. He found that given twice the time, a ball would roll four times as far. This result did not change when he made the slope gentler or steeper, so with a giant leap of imagination he assumed that it would still be true if the slope were vertical—a true falling body. To that he added mathematical analysis: if the distance was proportional to the square of the time, that meant, as he showed by geometric arguments, that the motion was uniformly accelerated. Finally, he imagined the body falling in a vacuum. In Aristotelian mechanics, a place is where something is. To imagine a place where there is nothing—a vacuum—is a contradiction in terms, an unthinkable logical absurdity. But Galileo broke free of at least some of the tethers of Aristotelian thought. He imagined a vacuum and realized that in a vacuum the acceleration of a falling body would not depend on how heavy the body was; only air resistance makes lighter bodies fall slower than heavy ones. That completed his law of falling bodies.

It did not, however, explain why a body would fall to the foot of the Tower of Pisa instead of landing half a mile away. Nevertheless, the answer to that dilemma, too, came out of the experiments with balls and inclined planes. Galileo found that if a ball was allowed to roll down one plane and back up another, it would tend to keep rolling up the second plane until it had reached the same height it started at. If

Portrait of Galileo, from *Il Saggiatore*, 1623.

the slope of the second plane was steeper than the first, the ball would roll a shorter distance, and if the slope was gentler, the ball would roll farther, achieving in both cases the same altitude it began with. Today we understand this behavior to be a manifestation of what we call the conservation of energy. But Galileo saw in it something else. Making another leap of the imagination, he reasoned that *if the second plane were horizontal, the ball would never stop rolling*, because it would never reach its original height. Thus, he concluded, the natural state of an object in horizontal motion was to keep on moving horizontally, at constant speed, forever.

This idea was a radical departure from Aristotelian philosophy, in which any horizontal motion required a proximate cause. It would eventually be transformed into Newton's first law of motion, the law of inertia. It was also exactly what was needed to resolve the dilemma of the object dropped from the Tower of Pisa—and, indeed, the more general question of why we don't sense the motion of the Earth. The surface of the Earth and everything upon it are all in horizontal motion together, due to the Earth's rotation. Their natural state is to continue that way, so to an observer on Earth's surface, everything, moving together, seems to be at rest. If the experiment at the Tower of Pisa were viewed by an observer truly at rest, it would be seen that both the tower and the falling object naturally move together in the horizontal direction, even as the object falls. Thus the object lands at the foot of the tower.

The same reasoning applied, said Galileo, to any projectile—for example, a cannonball. In the horizontal direction, a cannonball (ignoring air resistance) would retain the initial speed given it by the exploding gunpowder. Meanwhile, in the vertical direction, his law of falling bodies would apply, even while the cannonball was on the upward part of its trajectory. Combining these two types of motion, and unleashing his mathematics, Galileo showed that the path of any projectile near the surface of the Earth was a parabola. "It has been observed," he wrote in 1638 in *Two New Sciences*, "that . . . projectiles follow some kind of curved path, but that it is a parabola no one has shown. I will show that it is, together with other things, neither few in number nor less worth knowing, and, what I hold even more important, they open the door to a vast and crucial science." Once again, Galileo was correct: it was indeed to be a vast and crucial science. His discovery that inertia (the tendency of a body to keep moving at constant speed in the horizontal direction) combined with gravity (as represented by his law of falling bodies) produced trajectories near the Earth's surface in the form

DIALOGO
D I
GALILEO GALILEI LINCEO
MATEMATICO SOPRAORDINARIO

DELLO STVDIO DI PISA.

E Filofofo, e Matematico primario del

SERENISSIMO

GR.DVCA DI TOSCANA.

Doue ne i congreffi di quattro giornate fi difcorre
fopra i due

MASSIMI SISTEMI DEL MONDO
TOLEMAICO, E COPERNICANO;

*Proponendo indeterminatamente le ragioni Filofofiche, e Naturali
tanto per l'vna, quanto per l'altra parte.*

CON PRI VILEGI.

IN FIORENZA, Per Gio:Batifta Landini MDCXXXII.

CON LICENZA DE' SVPERIORI.

Title page, *Dialogo . . . sopra i due massimi sistemi del mondo, Tolemaico, e Copernicano*, 1632. For defending the Copernican theory in this book, Galileo was brought to trial before the Roman Inquisition and sentenced to perpetual house arrest. The book remained on the index of forbidden books until 1823.

of one of the conic sections, the parabola, was the very idea that Isaac Newton would later use to show how the universe works.

Galileo's eventual troubles with the Church—an epic tale, but not the subject of this book—had the effect of expelling the Scientific Revolution from Italy. It would come to rest in England, in the person of Isaac Newton. On the way north, however, it stopped briefly in France, where it found René Descartes. Descartes understood straight lines. In fact, the familiar x-y-z system of Cartesian coordinates is named for him. Galileo's version of inertia worked only in the horizontal direction. But when it is extended globally, horizontal motion at constant speed becomes circular motion about the center of the Earth. For all his nimble wit, Galileo could not quite escape this one remaining Platonic ideal. Descartes succeeded in straightening it all out. He put the law of inertia in the form used by Newton: in the absence of any external

René Descartes.

force acting on it, a body at rest will remain at rest, and a body in motion will remain in motion, at constant speed, in a straight line.

Isaac Newton is generally supposed to have been born in 1642, the year of Galileo's death, as if it were required that one such genius be present on Earth at all times. In reality, he was born on January 4, 1643, according to our modern calendar and the one then used in Galileo's Italy. In England, because of King Henry VIII's marital (or perhaps conceptual) difficulties, the latest papal calendar reform had not yet been adopted, and the date was rendered December 25, 1642. In any case, Newton was born both posthumously and prematurely, an unusual combination. His father (also Isaac Newton) died three months earlier, and the new Isaac was a frail creature, who did not seem destined to live on for eighty-four years.

Isaac's mother hoped he would grow up to manage the considerable properties left her by her second husband, who died when Isaac was about eleven years old. In fact, if Isaac's father had lived, or if his stepfather had been a more sympathetic person, Isaac might have grown up to be a reasonably well-adjusted, extremely bright farmer. But such a fate was not to be. Instead he grew up to be a man whose towering rage sometimes edged over into outright insanity, and who at the end of his life proclaimed himself to have remained a virgin. But he was also a man who would change human history as few others have done.

In 1661, the young Isaac matriculated at Trinity College, Cambridge, where Aristotle still commanded the curriculum but where the Scientific Revolution was in the air. Newton received his bachelor's degree in 1665, then fled to the family lands in Lincolnshire to escape the bubonic plague. It is thought that he made many of his most important discoveries during the two years he spent there, but the world would not hear of them until much later.

Among Newton's abundant accomplishments, the most important was to formulate a set of dynamical principles that would replace the Aristotelian worldview. By 1687, when he published his masterwork, the *Principia*, he had reduced it all to three laws, augmented by a number of definitions and corollaries. The first law was the principle of inertia, inherited from Galileo and Descartes:

LAW 1

Every body continues in its state of rest, or of uniform motion in a straight line, unless it is compelled to change that state by forces impressed upon it.

Newton's second law, the real centerpiece of his dynamics, tells what happens to a body when forces are indeed impressed upon it:

LAW 2

The change in motion is proportional to the motive force impressed; and is made in the direction of the straight line in which that force is impressed.

Earlier in the *Principia*, Newton defined quantity of motion to be the product of velocity (that is, speed plus direction) and quantity of matter—or precisely what today's physicists would call momentum. Long after Newton's death, his second law would come to be summarized in the equation $F = ma$ (force equals mass times acceleration); however, Newton never expressed it quite that way.

Newton's third law is called the law of action and reaction:

LAW 3

To every action there is always opposed an equal reaction: or, the mutual actions of two bodies upon each other are always equal, and directed to contrary parts.

The third law helps to eliminate a potentially messy complication in the problem of planetary motion. The planets (including the Earth) are great big complicated bodies, whose internal parts apply forces to one another. According to Newton's third law, these forces all cancel one another out, regardless of their nature. Every force due to one bit of a planet acting on a second bit is exactly balanced by an equal and opposite force due to the second bit acting on the first. The net result is that the bulk nature of the planet can be ignored completely in calculating its course around the Sun. The planet behaves exactly as if its mass were concentrated at a geometrical point located at its center.

The third law also implies that the planets impress forces upon the Sun equal and opposite to the Sun's forces on the planets. To get around any difficulties this might cause, Newton in his formal proofs refers not to the Sun but to "an immovable centre of force." In effect, he is assuming (correctly) that the Sun is so massive that it is not much influenced by the tug of gravitational forces from the planets. The third law would later prove vitally important in other areas of physics: it is the source of the laws of conservation of momentum, angular momentum, and energy. For the problem of planetary motion, however, its principal virtue is that all its effects may be ignored.

Isaac Newton, engraving by B. Reading, 1799, after a portrait by Sir Peter Lely.

Newton's three laws are the dynamical principles that replace the "natural motions" and "violent motions" of Aristotelian mechanics. To these laws, which apply to all forces and all bodies, Newton added the specific nature of a particular kind of force acting between the Sun and the planets, or between a planet and its moons—or, indeed, between any two bits of matter in the universe. This was the force of gravity, and as we shall see, he used Kepler's second and third laws to deduce the properties of the force of gravity. Then he demonstrated that his three laws, combined with the force of gravity, gave rise to elliptical orbits for the planets.

Isaac Newton invented differential and integral calculus. There is little doubt that he used these powerful analytic tools to make his great discoveries. When he wrote the *Principia*, however, he had not yet published his calculus. (There would later be a typically nasty dispute over priority with the German philosopher and mathematician Gottfried Leibniz, who made the same mathematical discoveries independently.) The *Principia* is presented in the classical languages of Latin and Euclidean geometry. The reason is obvious enough: Newton had to speak to his contemporaries in a language they would understand. There may have been another advantage to this method of presentation. Many years later, Richard Feynman (a man as different from Newton as history has yet produced, except for matters scientific) was intrigued enough to invent his own purely geometric proof of the law of elliptic orbits. "It is not easy to use the geometrical method to discover things," he said in his lecture on the subject (Chapter 4 of this book), "but the elegance of the demonstrations after the discoveries are made is really very great."

Isaac Newton is famously quoted as saying, "If I have seen further it is by standing on the shoulders of Giants." The giants were Copernicus, Brahe, Kepler, Galileo, and Descartes. Before Newton, the collapse of the Aristotelian worldview had left in its wake only buzzing confusion, without the remotest hint of how it might be replaced. Each of Newton's giants put in place a bit of the building material or some piece of the scaffolding, but the shape and design of the final structure could not be seen. (Descartes thought he saw it, but he was wrong.) Then along came Newton, and suddenly the world was orderly, predictable, and comprehensible once more. Newton had figured out how it all worked, and the proof that he was right was his demonstration of Kepler's law of ellipses. Soon we will do our own demonstration of the law of ellipses—not quite as Newton proved it but as Richard Feynman did almost three hundred years later.

First, a look at Richard Feynman.

2

Feynman: A Reminiscence

In 1965, when Richard Feynman shared the Nobel Prize with Julian Schwinger and Shinichiro Tomonaga for the invention of quantum electrodynamics, he was unknown to the general public, but he was already a hero of legendary proportions among physicists. At the time, the authors of this book were both graduate students at the University of Washington in Seattle, a lovely campus that seemed very far from the center of the intellectual universe. Nevertheless, early in 1966, when I (D.L.G.) started seriously looking for my first job, Caltech had an opening in experimental low-temperature physics. I was invited to come to Pasadena and give a seminar.

These were heady times in low-temperature physics. Low-temperature physics, the study of the behavior of matter at temperatures just above the unattainable absolute zero, was a coherent discipline rather than a mere set of techniques, because it had been organized around two central unsolved problems of many years' standing: superfluidity and superconductivity. Superfluidity is the mysterious ability of liquid helium to flow with no resistance at temperatures within two degrees of absolute zero. Superconductivity is the analogous ability of many metals to conduct electric current with no resistance at similar low temperatures. These phenomena had been unexplained for decades.

Then, in the 1950s, both problems were broken wide open, due in no small measure to Feynman's efforts. An intense period of creativity in both fields followed. For example, the new understanding of superconductivity made it possible to imagine ordinary electric circuits designed to use quantum mechanical devices. The most promising of these would be based on experiments by James Mercereau, a Caltech Ph.D. who had developed the Superconducting Quantum Interference Device, known universally in physics as the SQUID.

Feynman followed Mercereau's experiments avidly, and in fact could often be found in those days in Caltech's low-temperature physics laboratory, partly because of his intense interest in the experiments being performed there and partly because the low-temperature group had an extremely attractive secretary (who later became Mrs. Mercereau).

Under the circumstances, to be asked to fly out of the Seattle drizzle into the Pasadena sunshine to give a seminar to the low-temperature physics group was an offer I found irresistible. And Caltech had a few more tricks up its sleeve. Mercereau, who was seeking to strengthen Caltech's experimental low-temperature physics efforts, met me himself at the airport, and asked me whether I minded going to lunch before checking in at Caltech, mentioning that he had arranged for Dick Feynman to join us. The lunch with Feynman and Mercereau took place at a topless restaurant in Pasadena, Feynman's favorite hangout at the time. The only thing I can remember about that hour of culture shock is thinking to myself over and over again, "No one in Seattle is going to believe this." I had recovered well enough by the time I was due to give the seminar, and as it happened, in a few months we would come to Caltech to stay.

Richard Feynman was born on May 11, 1918, to Lucille and Melville Feynman. The powerful streetwise accent that he retained, and even honed, throughout his life suggested to most listeners that he was a native of Brooklyn, but in fact he was born and raised in Far Rockaway, in the sedate borough of Queens.

Feynman's father, whom Feynman revered in later days, was not very well off, but young Richard was recognized early as a prodigy, and so it was arranged for him to go to MIT, where he got his bachelor of science degree in 1939, and then to Princeton for his Ph.D. At Princeton, where his thesis adviser was John Archibald Wheeler, he worked on applying the principle of least action to quantum mechanics. His thesis laid the groundwork for some of the most important accomplishments of his later life.

During his graduate school days at Princeton, Feynman had his one and only encounter with Albert Einstein. Einstein was in Princeton at the Institute for Advanced Study, an establishment quite separate from the university. Nevertheless, members of the institute and members of the university's Physics Department often attended one another's seminars.

One day it was announced that the graduate student Richard Feynman would be presenting his first seminar. Not only was it to be his first seminar but he was to present and defend the startling idea that he and Wheeler had been working on: that an electron could move both forward and backward in time. Word got around that Einstein and a number of other famous physicists who happened to be visiting would attend.

Understandably nervous, the young Feynman decided to skip the usual pre-seminar tea and cookies, and instead prepared for his talk by going to the seminar room and filling the blackboard with equations. At a certain moment during this exercise, he sensed that someone was watching him. He turned around to see Albert Einstein in the doorway. The two great physicists looked at each other briefly, and then had the only private verbal exchange that ever occurred between them: "Young man," Einstein said, "where are they serving tea?" Years later, Feynman did not remember exactly what he said in reply.

While still at Princeton, Feynman married the girl of his dreams, Arlene Greenbaum. By the time he received his Ph.D., in 1942, the nation was at war. The young couple set out for Los Alamos, New Mexico, where a supersecret project to build an atomic bomb was being organized. Feynman joined the Theoretical Division at Los Alamos under the leadership of Hans Bethe, the great theorist who had figured out how the Sun and the stars burn their nuclear fuel. Arlene, who was already dying of tuberculosis, entered the hospital in Albuquerque.

During his days at Los Alamos, it became clear that Feynman could compete on equal terms with the intellectual giants of his day, including

Bethe, Enrico Fermi, and John von Neumann. At the same time, characteristics that would eventually become part of his legend surfaced: he showed his taste for mischief, cracking safes with simple tricks and planting teasing notes inside, and exchanging letters with his wife that were cut up like a jigsaw puzzle, so that the censors would have to spend a lot of time reconstructing them.

One day rather late in the game, Feynman had lunch with the patent officer for the Los Alamos project. Even though every aspect of the project, including its very existence, was strictly secret, this officer's job was to patent any new inventions that came up—probably in order to reserve the government's right to use them. Much to the patent officer's dismay, however, the scientists seemed to have little time and less inclination to seek patents. Come on, the officer complained to Feynman, you guys are creating a whole new world! Surely there must be some new things that can be done with it! Feynman thought briefly and said that he guessed there were—you could have an atomic submarine, for example, or an atomic airplane.

The morning after this feckless lunch, Feynman found on his desk complete patent applications, waiting for his signature, for the "Atomic Submarine" and the "Atomic Airplane." This is how Feynman came to hold the patent on the nuclear submarine, a device of considerable military significance but little commercial value. It has been said that years later, when Hughes Aircraft considered developing a nuclear-powered atomic airplane, they offered Feynman a vice-presidency (which he promptly turned down) because he held the patent. In any case, according to the patent agreement that Los Alamos workers signed, Feynman was entitled to one dollar for each patent. When he demanded his two dollars, it turned out that no fund had been set up for this purpose, so the patent officer was forced to pay the sum out of his own pocket. Feynman used the money at the canteen, to buy oranges and chocolates for everyone in the Theoretical Division.

In 1945, Arlene died in the hospital in Albuquerque, an episode Feynman wrote about movingly many years later, in *"What Do You Care What Other People Think?"* Feynman, who had borrowed his roommate's car to be at her bedside, returned to Los Alamos so despondent that he could not face the immediate prospect of talking about his wife's death. Instead his roommate arranged for the two of them to spend a quiet evening with some friends, who were not told what had happened that day. Years later, Feynman would recall marveling to himself that evening that others could remain oblivious of the enormous secret inside his head. His roommate was Klaus Fuchs, who had a few secrets of his own and was later convicted as a spy for the Soviet Union.

After the war, Feynman accepted from Hans Bethe a position at Cornell University, where he turned his attention to the quantum mechanical description of the interaction between light and matter. Although Schwinger and Tomonaga, who developed equivalent solutions to the problem independently, would share the Nobel Prize with him for this work, Feynman's approach was by far the most original. His technique discarded the electromagnetic field of James Clerk Maxwell and replaced it entirely with interactions among particles, following all possible paths with probabilities governed by the principle of least action, as adumbrated in his doctoral thesis. (In Chapter 3, we will see an echo of this approach to physics, when Feynman uses a kind of least-action principle as part of his geometric proof of the law of ellipses.) He also invented a method of using pictorial representations to keep track of the complex calculations his approach required. These representations have come to be known universally as Feynman diagrams. Feynman's work amounted to nothing less than a reformulation of quantum mechanics itself. His diagrammatic method is widely used in many areas of theoretical physics.

In 1950, Feynman left Cornell to join the faculty of the California Institute of Technology, where, except for a year in Brazil (1951–52), he would spend the rest of his career. At Caltech, he turned his attention to the problem of superfluidity in liquid helium. The Russian theorist Lev Landau had shown that the ability of superfluid helium to flow without resistance was due to the fact that the liquid could take up energy from its surroundings only in certain very restricted ways. Feynman succeeded in tracing Landau's observation to its quantum mechanical roots. Feynman diagrams would later become an important research tool in this field, but Feynman did not use them to solve this problem. Instead, he reverted to the old-fashioned, Schrödinger formulation of quantum mechanics and used his remarkable intuition to guess the nature of a giant quantum system.

During this period, Feynman's private notes show that he also tried very hard to solve the companion problem of superconductivity. The problem seemed ideally suited for Feynman's talents. As in the case of superfluidity, the solution would involve a gap in the energies that an electric current could absorb from its surroundings. Furthermore, that gap would arise as a consequence of the interactions between electrons in the metal and sound waves, or phonons. That part of the problem is closely analogous to the interactions between electrons and light waves, or photons, which had been the basis of Feynman's theory of quantum electrodynamics. Therefore (and unlike the matter of superfluidity), the Feynman diagram techniques, of which Feynman was, of course, the supreme master, seemed perfectly suited to the work. Feynman's chief competitors, John Bardeen, Leon Cooper, and J. Robert Schrieffer, were keenly and gloomily aware of all this. However, as it turned out, Feynman's powerful techniques led him inevitably in a direction that could not succeed, and it was Bardeen, Cooper, and Schrieffer who, early in 1957, found a dramatic solution to the problem. For their efforts, they won the Nobel Prize, Bardeen for the second time. (His first was in 1956 with William Shockley and Walter Brattain, for the transistor.)

Superconductivity was not the only problem Feynman tried but failed to conquer. In the course of his life, he also made forays into such arenas as experimental biology, statistical mechanics, Mayan hieroglyphics, and the physics of computing machines, with varying degrees of success. He was extremely reluctant to advertise or publish results in which he did not have complete confidence, or that might have stolen credit from deserving rivals; his list of publications is thus not long, and they are almost never wrong.

Shortly after arriving at Caltech, Feynman was joined there by Murray Gell-Mann, who was later to win his own Nobel Prize (1969) for revealing symmetries in the properties of the elementary particles of matter. With Feynman and Gell-Mann in residence, Caltech became the center of the universe of theoretical physics. In 1958, they published jointly a paper entitled "Theory of the Fermi Interaction"—an explanation of what has come to be known as the weak interaction, a fundamental

Feynman and Gell-Mann, 1959.

force that governs the decay of certain nuclear particles. Feynman and Gell-Mann were aware at the time that their theory was contradicted by experiments, but they had sufficient self-confidence to publish it anyway. The experiments later turned out to be mistaken: the theory was correct.

Also during this period, Feynman contributed to the work by Gell-Mann and by George Zweig, another Caltech professor of theoretical physics, that produced the theory of quarks, which is central to our present ideas of the nature of matter.

In 1952, Feynman married Mary Louise Bell, a university instructor in the history of decorative art. They were divorced in 1956. He was married for the third and final time on September 24, 1960, to Gweneth

Howarth. Their son, Carl, was born in 1962, and in 1968 they adopted a daughter, Michelle. Feynman cultivated a public persona—well known among his colleagues—that featured sketching nude women and spending time in topless bars, but his private life was solidly conventional and middle class, played out in a comfortable home in Altadena at the base of the San Gabriel Mountains, not far from the Caltech campus.

In 1961, Feynman undertook a project that would have far-reaching impact on the entire scientific community. He agreed to teach the two-year sequence of introductory physics courses that were required of all incoming Caltech students. His lectures were recorded and transcribed, and all the blackboards he filled with equations and sketches were photographed. From this material, his colleagues Robert Leighton and Matthew Sands, with help from Rochus Vogt, Gerry Neugebauer, and others, produced a series of books called *The Feynman Lectures on Physics*, which have become genuine, enduring classics of the scientific literature.

Feynman was a truly great teacher. He prided himself on being able to devise ways to explain even the most profound ideas to beginning students. Once, I said to him, "Dick, explain to me, so that I can understand it, why spin one-half particles obey Fermi-Dirac statistics." Sizing up his audience perfectly, Feynman said, "I'll prepare a freshman lecture on it." But he came back a few days later to say, "I couldn't do it. I couldn't reduce it to the freshman level. That means we don't really understand it."

Feynman delivered the *Feynman Lectures* to the Caltech freshman class in the academic year 1961–62 and to the same students as sophomores in 1962–63. His taste in physics topics was perfectly eclectic; he devoted just as much creative energy to describing the flow of water as to discussing curved spacetime. Of all the subjects he covered in that introductory course, perhaps his most impressive accomplishment is the presentation of quantum mechanics (Volume III of the series); in only slightly disguised form, it is the new view of quantum mechanics that he himself had developed.

While Feynman was a riveting, dramatic performer in the classroom, the period 1961-1963 was one of the few times he taught formal undergraduate courses. For the rest of his professional life, before and after, he taught mainly courses designed for graduate students. The lecture that is the subject of this book was not part of the original course but rather a "guest lecture" to the freshman class at the end of the winter quarter in 1964. Rochus Vogt had taken over the teaching of introductory physics by then, and he invited Feynman to give the talk as a treat for the students. The *Feynman Lectures* were never successful as introductory textbooks—not even at Caltech, where they originated. They would instead make their lasting contribution as a source of insight and inspiration for accomplished scientists who had learned their physics by more conventional means.

In the immediate aftermath of his Nobel Prize in 1965, Feynman suffered a brief period of dejection, during which he doubted his ability to continue to make useful, original contributions at the forefront of theoretical physics. It was during this time that I joined the Caltech faculty. The Feynman physics course was now being taught by Gerry Neugebauer. When Feynman himself had been giving the lectures, Gerry, as a young assistant professor, had had the difficult job of making up homework assignments from them for the two hundred or so students—difficult in large part because no one, maybe not even Feynman himself, knew in advance exactly what he was going to say. Just as he did for the lost lecture of Chapter 4 in this book, Feynman would come to class with no more preparation than one or two pages of scribbled notes. Neugebauer, to make his own task somewhat easier, would join Feynman, Leighton, and Sands for lunch after each lecture, in the Caltech cafeteria, known to generations of students as "the Greasy"; Caltech's elegant faculty club, the Athenaeum, was not Feynman's style. During these lunches, the lecture would be rehashed, with Leighton and Sands competing to score points with Feynman, while Neugebauer desperately tried to figure out the essence of the lecture.

Now, in 1966, Neugebauer was giving the lectures, and I was pressed into service as a T.A. (teaching assistant), in charge of one of the small recitation sections that supplemented the main course of lectures. The by now traditional lunches at the Greasy continued, with Feynman still in attendance. It was here that I first really got to know him, mostly exchanging ideas with him on how to teach physics. That fall, he got an invitation to give a public lecture at the University of Chicago the

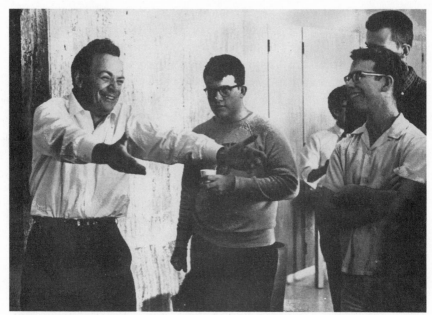

ABOVE. "I don't want to open it right away, so I fool around awhile."
Feynman telling Caltech students how he cracked safes at Los Alamos, 1964.

BELOW. Feynman and Leighton, 1962.

LEFT.
Feynman at the
blackboard,
1961.

BELOW.
Feynman on the
motion of
waves,
1962.

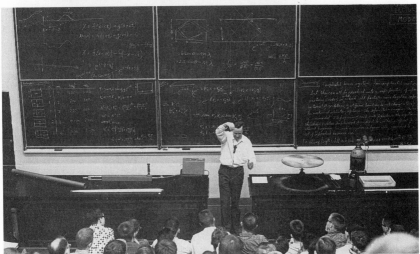

following February. At first he was inclined to refuse (invitations to speak arrived almost daily), but then he decided to accept and to talk about our ideas on teaching, if I would agree to come with him. He said that he would pay for my travel expenses out of the absurdly large ($1000) honorarium they were offering. I thought the matter over carefully for a microsecond or so, and agreed to go. When he told the University of Chicago that I would be joining him, they were no doubt mystified about who I was and why I was needed, but they invited me with good grace and paid my way in the bargain.

At Chicago, Feynman and I shared a suite in the Quadrangle Club, the university's faculty club. On the evening after his talk, we had dinner at the home of friends, Val and Lia Telegdi. The next morning, I wandered down to the faculty club dining room for breakfast a bit late. Feynman was already there, eating with someone I didn't know. I joined them, introductions were mumbled but not heard, and I sleepily drank my morning coffee. As I listened to the conversation, it dawned on me that this person was James Watson, discoverer with Francis Crick of the double-helical structure of DNA. He had with him a typed manuscript entitled *Honest Jim* (the title would later be changed by the publisher to *The Double Helix*), which he wanted Feynman to read, in the hope that Feynman might contribute something to the dust jacket. Feynman agreed to look at the manuscript.

That evening there was a cocktail party and dinner in Feynman's honor at the Quadrangle Club. At the cocktail party, the worried host asked me why Feynman wasn't there. I went up to the suite and found him immersed in Watson's manuscript. I insisted that since he was the honoree, he had to come down to the party. Reluctantly, he did, but he fled after dinner at the earliest moment permitted by civility. When the party broke up, I went back up to the suite. Feynman was waiting for me in the living room. "You've gotta read this book," he said.

"Sure," I said, "I'll look forward to it."

"No," he shot back, "I mean right now." And so, sitting in the living room of our suite, from one to five in the morning, with Feynman waiting impatiently for me to finish, I read the manuscript that would become *The Double Helix*. At a certain point, I looked up and said, "Dick, this guy must be either very smart or very lucky. He constantly claims he knew less about what was going on than anyone else in the field, but he still made the crucial discovery.'' Feynman virtually dove across the room to show me the notepad on which he'd been anxiously doodling while I read. There he had written one word, which he had proceeded to illuminate with drawings, as if he were working on some elaborate medieval manuscript. The word was ''Disregard!''

''That's what I'd forgotten!'' he shouted (in the middle of the night). ''You have to worry about your own work and ignore what everyone else is doing.'' At first light, he called his wife, Gweneth, and said, ''I think I've figured it out. Now I'll be able to work again!''

By the end of the 1960s, Feynman was back in action, consumed with the problem that would occupy his attention for the next decade or more. Collisions at extreme high energy of heavy particles such as neutrons and protons could be described entirely in terms of interactions of their internal parts. This was the ''parton'' theory, the internal parts being the quarks that his colleagues Murray Gell-Mann and George Zweig had earlier proposed, to which were later added particles known as gluons and so called because their role was to ''glue'' the quarks together. This model had such impressive success in predicting the results of experiments in high-energy particle accelerators that quark theory has come to be universally accepted among physicists, even though it has proved impossible to extract a quark from inside a proton or neutron for separate study.

Feynman's sense of humor was as singular as everything else about the man. In 1974, the world of physics was set on its ear by the nearly simultaneous discovery—at the Stanford Linear Accelerator (SLAC) and the Brookhaven National Laboratory, on Long Island—of a new particle. Called the J particle by the Brookhaven group and the ψ (psi) particle by the SLAC group, it quickly came to be known as the J/ψ particle. The discovery was in the form of two very narrow peaks, called ''resonances,'' in a plot of detector signal versus accelerator-beam energy. At all other accelerator-beam energies, the detectors registered only a meaningless, low-level background noise. At the time, I was chairman of the Caltech Physics Department's colloquium committee. Since I was known to be a friend of Feynman's, the committee prevailed upon me to ask Dick to give a colloquium explaining the meaning of these stunning new discoveries. Feynman agreed immediately, and outlined for me the kind of talk he had in mind to give. We penciled in the earliest available date—January 16, 1975—and let it go at that. Having filled that date on my colloquium calendar, I put it out of mind as a fait accompli.

Three weeks before the appointed date, during the Christmas vacation, the editor of the weekly *Caltech Calendar* called me up. The title of Professor Feynman's colloquium was due immediately for publication in the calendar. Feynman was away at a family retreat in Baja California that, quite purposely, had no telephone. I had a big problem.

I invented a title for Feynman's talk. It was "The Broad Theoretical Background of Two Narrow Resonances." To a physicist it was a mild play on words; to anyone else it was incomprehensible. But it did describe perfectly the talk Feynman was planning. I called a mutual friend, Jon Mathews, to ask his advice. Jon laughed when he heard my title, but then sobered instantly and said, "Don't do it. Dick has a wonderful sense of humor about everything else, but he has no sense of humor at all about physics."

But I really liked my title, and it had made Jon laugh. I called it in to the calendar editor and promptly forgot the whole matter.

Feynman's colloquium was to be the second of the new year. On the day of the preceding one—Thursday, January 9—when we gathered for tea at 4:45, I saw Feynman for the first time since the vacation, and it all came flooding back to me. I also realized that the next week's calendar was out that day and that Feynman would have seen my invented title. By now, I feared the worst, but I met the problem head-on. "Look, Dick, I'm sorry," I babbled, "I had to give them a title, and you weren't there, so I did the best I could."

He stared down his nose at me, in a way that only he could do. "It's all right," he said, in a tone of voice that let me know that the story was far from over. "It's all right," he repeated ominously.

After a few minutes of tea drinking, we all went upstairs to the hallowed hall in which Caltech physics colloquia have been held since time immemorial (1921). As he often did, Feynman sat down next to me in the first row—informally but rigidly reserved for professors of physics. The talk was theoretical, technical, and difficult: "Equilibration Processes in Nuclei," by Steven Koonin, then a graduate student at MIT (he is currently Caltech's provost). Throughout, Feynman whispered commentary and wisecracks in my ear, so that by the end of the talk I had lost the thread of Koonin's argument entirely.

At the conclusion of the talk, another first-row seat holder, the nuclear physicist Willy Fowler, asked a question. (Willy would, in 1983, win his own Nobel Prize for the work he did on the production of elements in stars.) Although I hadn't understood much of the talk, I thought I did understand Willy's question, and I thought I knew the answer. Turnabout being fair play, I whispered my answer into Feynman's ear. Feynman's hand shot up instantly.

To speakers at the Caltech physics colloquia in those days, the audience consisted of Richard Feynman and a blur of other, indistinguishable faces. When Feynman's hand went up, young Koonin, who had been struggling to formulate an answer to Willy's question, called on him in manifest relief. Feynman solemnly rose to his feet (something that was *never* done during the question-and-answer period after a colloquium). "Goodshtein says"—he intoned my name in a stentorian voice, purposely mispronouncing it to sound like the German pronunciation of "Einstein"—"Goodshtein says . . . ," and he proceeded to give my answer, not as I had mumbled it to him but beautifully, elegantly phrased, as I could not possibly have done.

"That's it!" Koonin exclaimed. "That's *exactly* what I've been trying to say!"

"Well," said Feynman, as I tried to slip under the seat, "don't ask *me*. I don't understand it. That's what *Goodshtein* says." He had had his revenge. The matter was never mentioned again.

On a Friday in early June 1979, Feynman's trusted secretary, Helen Tuck, quietly called me up to say that he had been told that he had stomach cancer. He was to go into the hospital for surgery at the end of the following week. It was not entirely certain that he would come out again. I was not to tell him that I knew.

That Friday was commencement day at Caltech. There was Feynman, robing up to march in the academic procession. I told him that someone had reported a mistake in some work we had done together but that I couldn't find the source of the error. Would he like to talk about it? We agreed to meet in my office Monday morning.

On Monday morning we got to work. Or, rather, he did. I looked over his shoulder, commented, kibitzed, and mostly marveled to myself that this man, who was facing in secret a possibly fatal operation, was working with unflagging energy on an unimportant problem in two-dimensional elastic theory. The solution to the problem could be found in standard textbooks, but that wasn't the point. When we did the original work together, Feynman had insisted on working out this minor result himself, on a napkin in a topless bar, and we had unwisely published the result (a small part of a much larger theory) without checking it against the standard formula. At that point in his life, in spite of the time spent in topless bars, Feynman never drank anything alcoholic, out of fear of diminishing his mental powers. He could not be accused of Deriving Under the Influence. Nevertheless, something had gone wrong. The question was, what subtle error had he made that led to a slightly wrong answer?

The problem proved intractable. At 6 P.M., we declared the situation hopeless and went our separate ways. Two hours later, he called me up at home. He had the answer! He had not been able to stop working on it, he told me excitedly, and had finally tracked it down. He dictated the solution to me. Four days before entering the hospital for his first cancer operation, Richard Feynman was bursting with joy.

The tumor that was removed from his body the following weekend was large but seemed to the doctors to be well encapsulated, and a hopeful prognosis was given. Nevertheless, he would eventually die of the disease.

In the 1980s, in the last decade of his life, Feynman became a genuine public figure, perhaps the best-known scientist since Albert Einstein. Earlier in his career, even while cultivating his very special image among scientists, he had largely shunned public notice. He had even briefly considered refusing to accept the Nobel Prize, until he realized that such a gesture would bring him more attention than the prize itself. However, at the end of his career, a number of events conspired to make Feynman a celebrity.

In 1985, *"Surely You're Joking, Mr. Feynman!"* became a surprise runaway best-seller. Tales that Feynman had been telling about himself for years had been collected by Feynman's bongo-drumming buddy Ralph Leighton and edited by Edward Hutchings, a longtime Caltech journalism instructor. Subtitled *Adventures of a Curious Character* (the double meaning was intentional), the book recounted Feynman's nonscientific adventures, from his amilitaristic antics at Los Alamos to dancing in the Carnival at Rio. This unconventional view of the great scientist at play enthralled a public that had no idea why Feynman deserved his fame. The book was followed three years later by a second volume, *"What Do You Care What Other People Think?,"* subtitled *Further Adventures of a Curious Character*, also "as told to" Ralph Leighton.

In the meantime, however, a disastrous event had captured the national attention. On January 28, 1986, the space shuttle *Challenger* exploded just a few moments after liftoff. Seen live by millions and repeated endlessly on television, the scene burned itself into the American consciousness. A few days later, the acting head of the National Aeronautics and Space Administration (NASA), William Graham, who had been a Caltech student and had attended the lectures that Feynman gave regularly at Hughes Aircraft Company, called Feynman to invite him to serve on a presidential commission to investigate the accident.

The commission was headed by former secretary of state William P. Rogers, who had considerable difficulty containing the ebullient scientist. Feynman did not permit his advancing illness to interfere with his self-appointed role as an aggressive investigator. The peak moment of his public fame came when, during a nationally televised hearing of the commission, he squeezed a bit of gasket material from one of the shuttle's solid-fuel booster rockets in a clamp and dropped the sample into a glass of ice water, to demonstrate that the gasket material lost its resiliency at freezing temperature. (Contrary to the electrifying impression he created at the time, the demonstration had been carefully rehearsed in advance.) This defect did indeed prove to be the principal cause of the *Challenger* disaster.

Feynman's lost lecture on planetary motion was by no means the only ad-hoc lecture he ever gave for the benefit of the Caltech undergraduates. Over the years, he was often asked to make a guest appearance, and he nearly always complied. The last of these guest lectures took place on Friday morning, March 13, 1987. I was now teaching the freshman introductory physics course, and he agreed to my request to give the final lecture of the fall quarter.

The subject of Feynman's lecture on this occasion was to be curved spacetime (Einstein's theory of general relativity). Before starting, however, he had a few words to say on a subject that excited him greatly. Three weeks earlier a supernova had occurred at the edge of our galaxy. "Tycho Brahe had his supernova," Feynman told the class, "and Kepler had his. Then there weren't any for four hundred years. Now I have *mine.*"

This remark was greeted with a stunned silence by the freshmen, who had reason enough to be in awe of Feynman even before he opened his mouth. Dick grinned with obvious pleasure at the effect he had created, and defused it in the next breath. "You know," he mused, "there are about a hundred billion stars in a galaxy—ten to the eleventh power. That used to be considered a huge number. We used to call numbers like that 'astronomical numbers.' Today it's less than the national debt. We ought to call them 'economical numbers.' " The class dissolved in laughter, and Feynman went on with his lecture.

Richard Feynman died eleven months later, on February 15, 1988.

3

Feynman's Proof of the Law of Ellipses

"Simple things have simple demonstrations," Feynman wrote in his lecture notes. Then he crossed out the second "simple" and replaced it with "elementary." The simple thing he had in mind was Kepler's first law, the law of ellipses. The demonstration he was about to present would indeed be elementary, in the sense that it used no mathematics more advanced than high school geometry, but it would be far from simple.

To begin with, Feynman reminds us that an ellipse is a kind of elongated circle that can be made with two tacks, a string, and a pencil, like this:

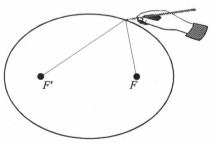

Each tack is at a point called a focus of the ellipse. The string makes a line from one focus to a point on the ellipse and back to the other focus. The total length of the string stays the same as the pencil goes around the curve. Here's a slightly more proper geometric diagram:

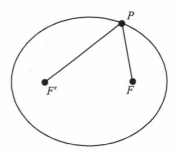

Here F' and F are the two foci, and P may be any point on the curve. The distance from F' to P and back to F is the same, no matter where P is on the curve.

Here is a small point worth remembering: If the string is made a little shorter and the tacks stay where they are, we get another ellipse, inside this one; and if the string is made longer while the tacks stay where they are, we get an ellipse that lies outside this one. It follows that any point in the plane—say, q—such that the distance from F' to q to F is less than the distance from F' to P to F (in other words, any point that can be reached by a shorter string) lies inside our original ellipse. Likewise, any point Q such that $F'Q + QF$ (another way of saying the distance from F' to Q plus the distance from Q to F) is larger than $F'P + PF$ (the length of the original string) lies outside our ellipse. Here's a picture illustrating the idea:

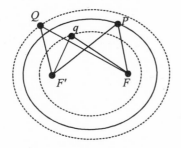

Any point Q outside the ellipse lies on a bigger ellipse, reached by a longer string. Any point q inside the ellipse lies on a smaller ellipse, reached by a shorter string.

A little later in the discussion, Feynman uses this idea, but he doesn't prove it as we have just done. Instead he asks the students to work it out for themselves.

An ellipse has another special property. If a lightbulb were turned on at F, and if the inner surface of the ellipse reflected light like a mirror, then all the reflected light rays would come back together at F', the other focus, like this:

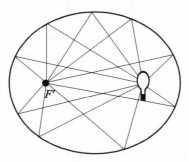

And vice versa: all the light rays starting at one focus will be focused to a point at the other focus. Feynman cites this as the second elementary property of the ellipse, and then he sets out to prove that these two properties are really equivalent. (His strategy here is to lead us to a more arcane property of ellipses—one that will prove indispensable later on.)

Picture any point P on the ellipse. At that point (as at any point) on the ellipse (or any other curve), there is a single, unique straight line that just touches the curve without penetrating it, like this:

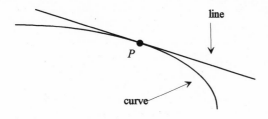

This line is called the tangent to the curve at that point. A light ray, mirror-reflected from the curve at any point, like this,

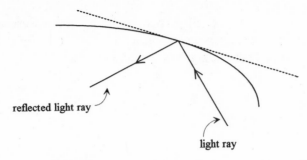

reflected light ray

light ray

follows the same path it would follow if it were mirror-reflected at that point from the tangent line, like this:

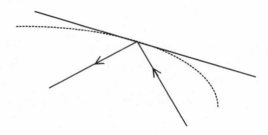

The reason that light reflects from the curve just as it would from the tangent line at the same point is that the tangent indicates the direction of the curve at exactly that point. If one starts with a curve and its tangent at a point,

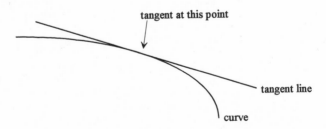

tangent at this point

tangent line

curve

and magnifies the picture greatly about that point, the curve is stretched out to become much more nearly equal to the tangent line:

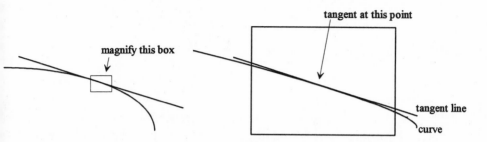

The more closely we look, the less difference there is between the curve and its tangent line at that point. Thus, if light is reflected from a curve at a single point, it reflects just as it would from the tangent line at that point. For the same reason, the tangent line has another property that will be important to us later on: if the curve is actually the path of a moving object, the tangent line shows the direction of the object's motion at each point. When we think of the ellipse as the path followed by a planet in its orbit around the Sun, the tangent to the ellipse at each point will be in the direction of the planet's instantaneous velocity at that point.

The law of reflection from a flat mirror is that the ray strikes the mirror and is reflected from it at the same angle, like this:

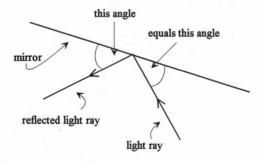

So here is the property for light rays:

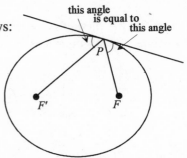

this angle is equal to this angle

The incident ray from F to P makes the same angle with the tangent line at P as does the reflected ray, which goes to F'. Our job is to prove that this statement is equivalent to saying that the distance $F'P$ plus the distance PF is the same for any point P on the curve.

The proof involves some new construction. A line is drawn from F' perpendicular to the tangent line, like this:

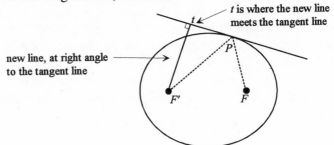

t is where the new line meets the tangent line

new line, at right angle to the tangent line

Then it is extended the same distance, to a point called G':

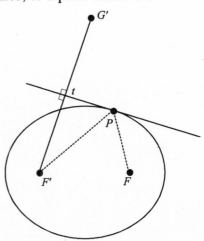

Thus the line $F'G'$ has been constructed in such a way that the line tangent to the ellipse at point P is its perpendicular bisector. Feynman calls G' the image point of F'. What he means is that if the tangent line were indeed a mirror, and if the point F' looked at itself in that mirror, its image would appear to be at G', an equal distance behind the mirror.

One more piece of construction is called for. Connect the points G' and P with a straight line:

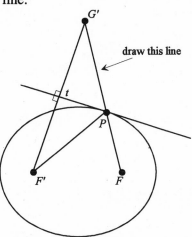

Now take a look at the two triangles that have been formed, one shown with solid lines and the other with broken lines:

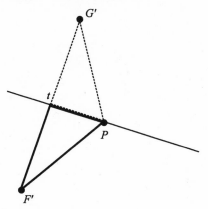

These two triangles are congruent, which means that they are identical in all respects except orientation. Here's the proof. Since we constructed the intersection at *t* to be a crossing of perpendicular lines, each triangle has one right angle:

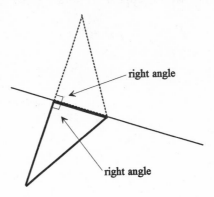

They share one side in common:

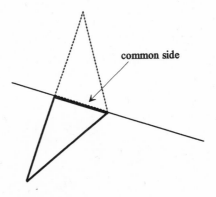

And another side of each was constructed to have equal length (remember, the tangent line bisects $F'G'$:

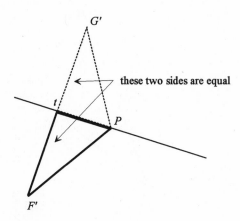

these two sides are equal

Any two triangles that have one equal angle and two equal sides are congruent; QED, as we used to say in high school. That means *all* the corresponding sides are equal. In particular, note that the side $G'P$ is equal to the side $F'P$:

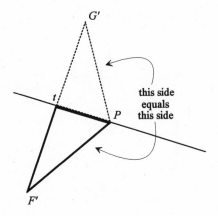

this side equals this side

And the angles $F'Pt$ and $G'Pt$ are equal:

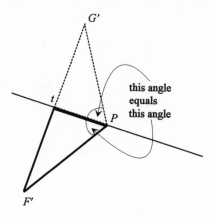

Okay, now back to the full diagram to see what we've learned.

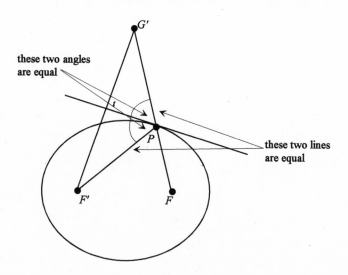

By this time, it's easy to have lost sight of what we're assuming and what we want to prove. To clarify the situation, let's reconstruct the same diagram from scratch. Start with two points, F' and F, that for the moment have no particular significance. They are any two points in a plane:

●
F'

●
F

Then draw any straight line from F' in any direction:

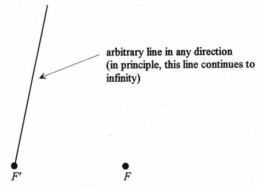

arbitrary line in any direction
(in principle, this line continues to infinity)

●
F'

●
F

Now pick a point t on the line and draw a perpendicular line through it. The point t must be far enough away from F' so that the perpendicular line doesn't pass between F and F':

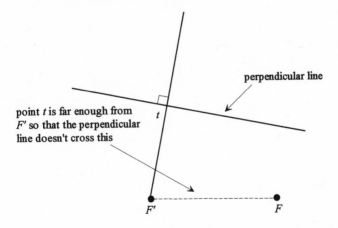

perpendicular line

point t is far enough from F' so that the perpendicular line doesn't cross this

t

●
F'

●
F

Mark a point G' on the arbitrary line, such that $F't$ is equal to tG'. Then the perpendicular we constructed is the perpendicular bisector of $F'G'$:

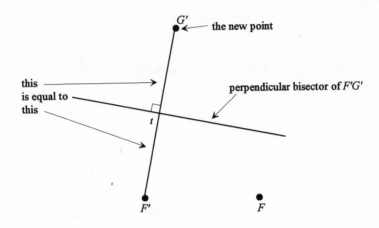

Next draw a line connecting G' and F:

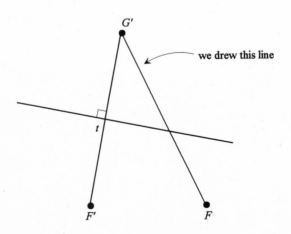

Label the point where this new line crosses the perpendicular bisector P, and draw the line connecting P to F':

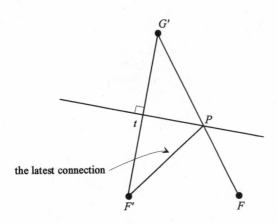

The two triangles are congruent as before, so the angles $F'Pt$ and $G'Pt$ are equal:

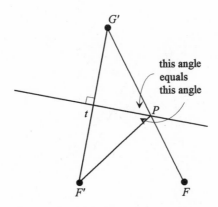

And the angle $G'Pt$ is also equal to the opposite angle where $G'PF$ crosses the perpendicular bisector (when two straight lines cross, the opposite angles are always equal):

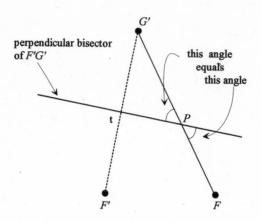

perpendicular bisector
of $F'G'$

this angle
equals
this angle

Therefore all these angles are equal:

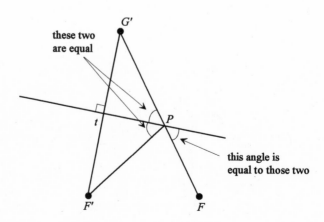

these two
are equal

this angle is
equal to those two

This means the perpendicular bisector line would reflect light from F to F' at the point P (because, at that point, the angles of incidence and reflection are equal). Not only that, the line FPG' has a really spectacular property, which can be seen by going back to the congruent triangles:

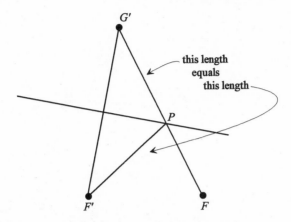

Because of the congruency of the triangles, the length $F'P$ is the same as the length $G'P$. It follows that the distance from F' to P and back to F is the same as the distance in a straight line from F to G'. But that distance is just the length of the string we used to draw our original ellipse. In other words, if we draw an ellipse by the string method, G' is the point we reach by straightening out the string:

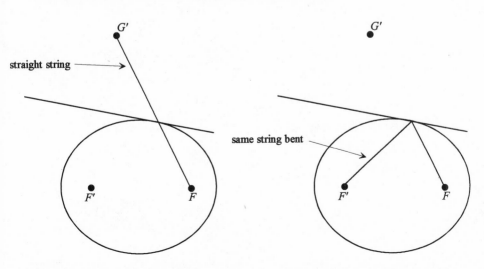

So we've discovered a strange and marvelous new way to construct an ellipse. Here's how it works. Take two points in a plane, F' and F. Then take a string of constant length (larger than the distance $F'F$) and connect one end to the point F. Stretch the string straight in any direction, mark the endpoint, and call it G':

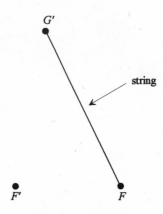

Next, connect F' and G', and draw the perpendicular bisector of $F'G'$. The perpendicular bisector crosses the line FG' at a point P:

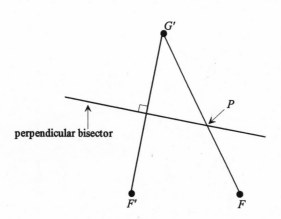

Now let the point G' at the end of the string move in a circle of constant radius centered at F:

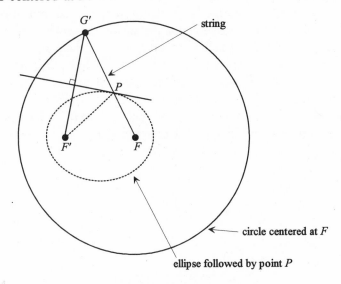

string

circle centered at F

ellipse followed by point P

As it does so, the point P, formed by the intersection of FG' and the perpendicular bisector of $F'G'$, traces out exactly the same ellipse that would have been formed using the same string with its ends tacked to F' and F! We know that, because we've proved that when P is constructed in this way, the distance FPG' (which goes from F to the circle) is always equal to the distance FPF' (which constructs the ellipse):

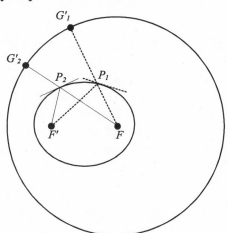

When G' moves from G'_1 to G'_2, P moves from P_1 to P_2, tracing out the ellipse

So, within every circle there lurks, for every off-center point, an off-center ellipse. However, while this is very interesting (and will later turn out to be very valuable), it's not the property we set out to prove.

What we did set out to prove is that the string-and-tacks construction of the ellipse is equivalent to its property of reflecting light rays from F to F'. What we have is an ellipse that obeys the string-and-tacks construction (that is, $F'P + PF$ is the same all the way around the ellipse), and the line that reflects light that arrives from F at point P, with equal angles of incidence and reflection, back to F'. That reflecting line happens to be the perpendicular bisector of $F'G'$:

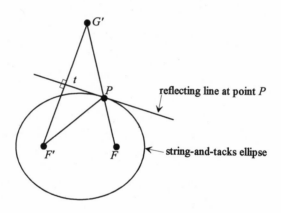

All that's left to prove is that the reflecting line at point P is also tangent to the ellipse at point P. We know that each point on the ellipse has the same mirror-reflection properties as a tangent line at that point. Thus, if the reflecting line at P is also tangent to the ellipse at P, then the ellipse reflects light from F to F' at any point P, and we have proved that the two properties (string-and-tacks and reflecting light from one focus to the other) are equivalent.

The proof is made by showing that while the point P is (by construction) on both the line and the ellipse, every other point on the line lies outside the ellipse. That is the unique property of the tangent to any curve at a point: it touches the curve without crossing it. If the line crossed the ellipse at P, part of the line would necessarily be inside it:

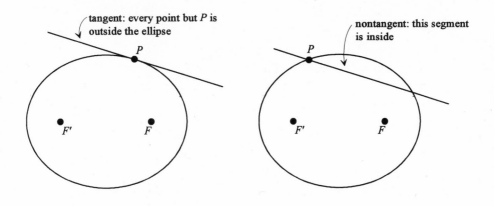

Go back to the construction and pick any point on the line other than P. Label that point Q, and connect it to F' and G':

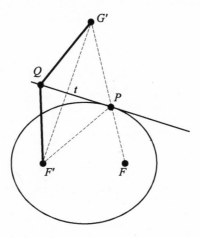

It should be easy to see now that the distances $F'Q$ and $G'Q$ are equal (PQ is the perpendicular bisector of $F'G'$, the triangles $F'tQ$ and $G'tQ$ are congruent, and so on, QED). Now draw the line QF:

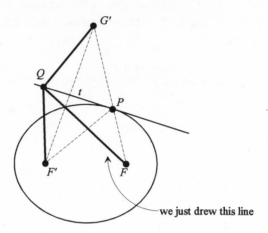

we just drew this line

The distance from F' to Q to F is equal to the distance from G' to Q to F; we know this because we know that the first steps are equal ($F'Q$ and $G'Q$) and the second steps are the same (QF). Now compare the lengths $FQ + QG'$ (solid lines) and $FP + PG'$ (broken line):

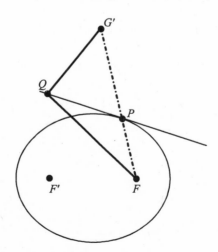

Obviously, *FPG'* is shorter, since it's a straight line, and a straight line is the shortest distance between two points. But we've just shown that the solid lines *G'QF* in the drawing above cover the same distance as the solid lines *F'QF* in the drawing below, and likewise for the broken lines (for the broken lines, we saw that earlier; it's the length of the string):

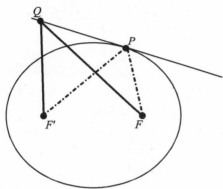

We have proved that the solid lines cover a bigger distance than the broken lines. In other words, if we wanted to reach point *Q* with a string stretched from tacks at *F'* and *F*, the string would have to be longer than the one needed to reach the unique point *P*. We showed much earlier that this means all such points are outside the ellipse. Thus, the line is tangent to the ellipse at point *P*. QED.

Speaking of QED, there's something particularly interesting in Feynman's use of this method of proof. We have shown in effect that the shortest path from *F'* to the tangent line and thence to *F* is the path that reflects light at point *P*. This is a special case of Fermat's principle (light always takes the quickest path between two points) and is closely related to Feynman's approach to quantum electrodynamics, which is also known as QED and won him his Nobel Prize. Fermat's principle is a special case of the principle of least action.

In any case, Feynman has now told us all we'll need to know about the ellipse. He turns now to dynamics—that is, forces and the motions that result from them. The diagram that Feynman has sketched in his lecture notes is copied directly out of Newton's *Principia*. That much is obvious from comparing them:

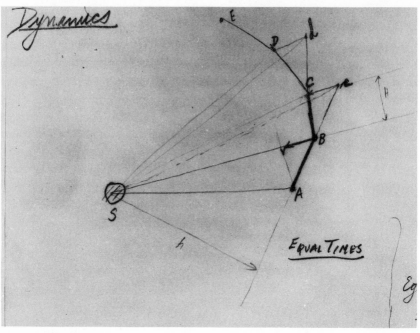

Feynman's Diagram

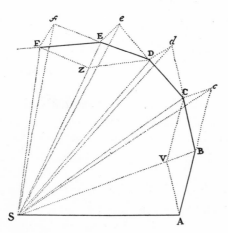

Newton's Diagram

In Newton's diagram, S represents the position of the Sun (the "immovable centre of force"), while A, B, C, D, E, and F are successive positions, at equal intervals of time, of a planet in orbit about the Sun. The motion of the planet is the result of a competition between the planet's tendency to move at constant speed in a straight line if no forces act upon it (the law of inertia) and the motion due to the force that is acting on the planet—that is, the gravitational force directed toward the Sun. In reality, these combined effects produce a smoothly curved orbit, but for purposes of seventeenth-century geometrical analysis, Newton represents them by a series of straight-line segments due to inertia, interrupted by sudden changes in direction due to impulsive (essentially instantaneous) applications of the Sun's force. Thus, the first bit of the diagram starts this way:

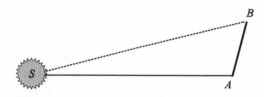

In a certain interval of time, the planet would move from *A* to *B*, if no force were acting on it. In the next equal interval of time, if there were no force acting, the planet would continue the straight line for an equal distance, *Bc*:

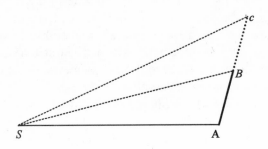

Instead, however, the Sun's force (which really acts continuously) is represented by an impulse applied at point *B*, which results in a component of motion directed toward the Sun, *BV*:

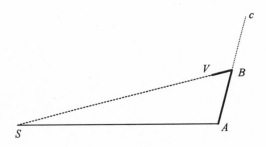

The motion the planet would have without the force, *Bc*, and the motion due to the force, *BV*, are compounded into a parallelogram; its diagonal is the "actual" motion:

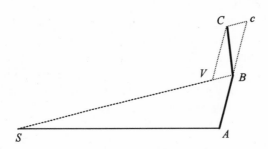

Thus, the planet "actually" follows the path *ABC*. Notice that *Cc* is not directed toward the Sun. It is strictly parallel to *VB*, which *is* directed toward the Sun. Incidentally, all of these points lie in a plane: any three points, *S*, *A*, *B*, define a plane. The lines connecting *S*, *A*, and *B* are in the plane. The segment *BV* lies in the same plane, because it's on the line *BS*. The segment *Bc* is in the plane, because it extends the line *AB*. The line *BC* is in the plane, because it's the diagonal of the parallelogram formed by *BV* and *Bc*. Now the same procedure is repeated at each point, so that the next step looks like this:

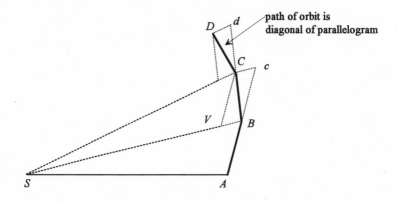

path of orbit is diagonal of parallelogram

And so on. In the end, Newton applies the same analysis to shorter and shorter equal time intervals, and the resulting path, *ABCD* . . . , becomes arbitrarily close to a smooth orbit, on which both inertia and the Sun's force act continuously. The orbit always lies in a single plane.

Before shrinking the time interval, Newton (and Feynman) now proves that the planet's orbit sweeps out equal areas in equal times. In other words, the triangle *SAB*, swept out by the planet in the first time interval, has the same area as the triangle *SBC*, swept out in the second equal time interval, and so on. The first step, however, is to show that triangle *SAB* has the same area as *SBc*—a triangle that would have been swept out in the second time interval if there were no force from the Sun. Here's what the three triangles look like:

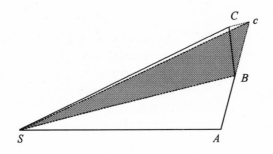

The area of a triangle is equal to one-half its base times its altitude. For example, one way to calculate the area of the triangle *SAB* would be to choose *SA* as the base, in which case the altitude would be the perpendicular distance from the continuation of *SA* to the highest point on the triangle:

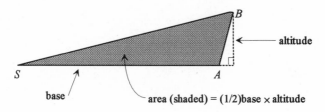

We get the same result if we choose *SB* as the base and construct the altitude like this:

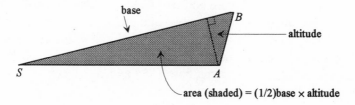

Now we want to compare that area to the area of *SBc*,

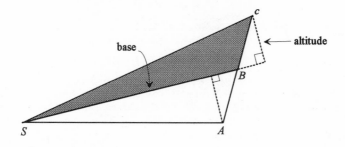

where we've chosen *SB* to be the base and constructed the altitude as shown. Look at the diagram formed by the construction of the altitudes of the two triangles:

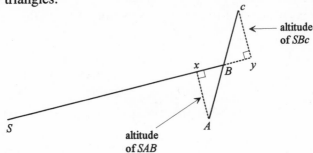

For the moment, the corners where right angles were constructed are labeled *x* and *y*. The triangles *ABx* and *cBy* are congruent, because they have one equal side and two equal angles. The equal sides are *AB* and *Bc* (equal because they are the distances the planet would go in equal time intervals if there were no force from the Sun), and the equal angles are the right angles (*AxB* and *Byc*) and the opposite angles made by the crossing of the two straight lines *xBy* and *ABc*. Since the triangles are congruent, the two altitudes, *Ax* and *cy*, are equal; and since the triangles *SAB* and *SBc* have the same base (*SB*) and equal altitudes, their areas are equal. QED.[1]

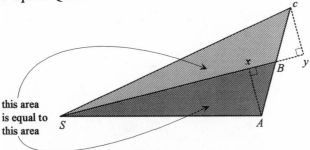

this area is equal to this area

[1]In Feynman's lecture, on page 155, where he does this proof, he chose *AB* and *Bc* as the bases of the two triangles. Then they both have the same altitude, formed by extending the line *ABc* downward, and constructing a perpendicular from it to *S*. This proof and the one in the text work equally well.

Next (following Newton and Feynman), we show that the area of *SBc* (solid lines) is also equal to the area of *SBC* (broken lines):

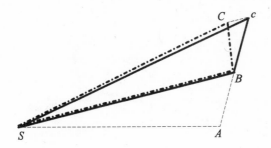

The two triangles have the same base, *SB*. The altitude of *SBC* is the perpendicular distance from the extension of *SB* to *C*:

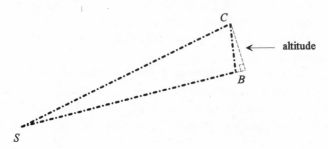

The altitude of *SBc* is the perpendicular distance from a farther extension of *SB* to *c*:

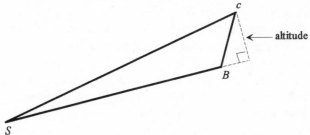

Put the two diagrams back together, and remember that Cc is strictly parallel to SB:

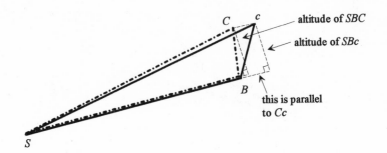

The two altitudes are the perpendicular distances between the same two parallel lines, and are therefore equal. Thus triangles SBC and SBc have the same base and equal altitudes. They therefore have the same areas. Once again, QED.

Aside from being very pretty geometry, this last proof is very important for physics. The path Bc would have been taken if there were no force at all. Instead, there is a force, directed toward S. That force changes the trajectory from path Bc to path BC, but it cannot change the area swept out during a fixed interval of time. In later years (after Newton but long before Feynman), this area would be understood to be proportional to a quantity called the *angular momentum*. In the language of latter-day physics, we have proved that a force on a planet directed toward S cannot change the angular momentum of the planet measured with respect to S. Although Newton never used the term "angular momentum," it is clear that he understood the significance of that quantity, and the fact that it could be changed only by a force along some direction not pointing at the center, S.

In any case, we have now shown that the area of SAB is equal to the area of SBc and that the area of SBc is equal to the area of SBC. It follows that SAB and SBC have the same areas. Looking back at the original diagram,

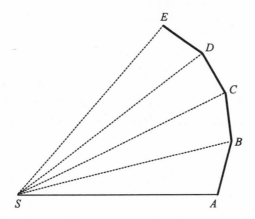

it is obvious that we could apply the same arguments to successive triangles—*SCD*, *SDE*, and so on. These are the triangles swept out by the planet in equal intervals of time. We have thus succeeded in proving Kepler's second law of planetary motion: a planet sweeps out equal areas in equal times.

Now that we can see where we have arrived, it is worthwhile to look back and see how we got here. What exactly did we have to know about dynamics—that is, about forces and the motions they produce—in order to get this far?

The answer is this: We have used Newton's first law (the law of inertia), Newton's second law (any change of motion is in the direction of the impressed force), and the idea that the force of gravity on the planet is directed toward the Sun. Nothing else. For example, we have not used the idea that the force of gravity is inversely proportional to the square of the distance. So the inverse-square-of-the-distance character of gravity has nothing to do with Kepler's second law. Any other kind of force would have produced the same result, provided only that the force is directed toward the Sun. What we have learned is that if Newton's first and second laws are correct, then Kepler's observation that planets sweep out equal areas in equal times means that the gravitational force on the planet is directed toward the Sun.

You may wonder exactly where we used Newton's first and second laws. We used the first law when we said the planet would move from A to B to c if there were no force on it, and the second when we said that the change in the motion, BV, due to the force from the Sun, is directed toward the Sun. Incidentally, we have also used Newton's first corollary to his laws— that the net motion produced by both tendencies in the time interval is given by the diagonal of the parallelogram of the separate motions that would have occurred:

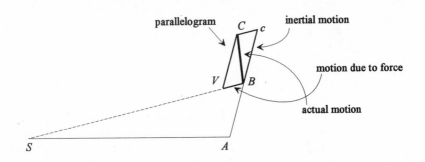

At this point in his lecture, Feynman says, "The demonstration that you have just seen is an exact copy of one in the *Principia Mathematica* by Newton," but he goes on to say that he could not follow Newton's arguments any further, and that he "cooked up" the rest of the demonstration of the law of ellipses himself. Before turning to Feynman's demonstration, however, let us interject another argument that Feynman has disposed of earlier in his lecture: where does the inverse-square-of-the-distance force of gravity come in?

The inverse-square-of-the-distance (from now on we'll just call it the R^{-2}) nature of gravity is deduced from Kepler's third law, which says that the time it takes a planet to make one complete orbit (that is, one year in the life of the planet) is proportional to the 3/2 power of the planet's distance from the Sun. Actually, since the orbits of the planets are ellipses with the Sun at one focus, a given planet is not always the same distance from the Sun:

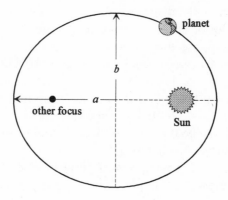

The distance from the center of the ellipse (*not* from the Sun, which is off-center) to the farthest point on the ellipse is called the semimajor axis, labeled *a* (the shorter axis, labeled *b*, is called the semiminor axis). The semimajor axis is called that because it is one-half the longest axis of the ellipse. Kepler's third law says that the time it takes a planet to execute one complete orbit is proportional to the 3/2 power of *a*, the semimajor axis.

Just to be sure the meaning of that statement is clear, imagine a sun with two planets in orbit around it (or a planet with two moons in orbit around it—the same law would be obeyed):

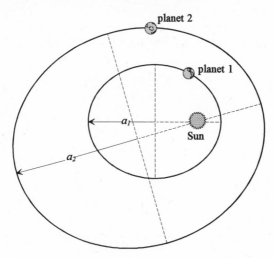

The two arrows show the distances from the centers of the two ellipses to the farthest point of each. Those distances are the semimajor axes, a_1 and a_2. Now suppose that a_2 is twice as big as a_1. Then Kepler's third law says that the time planet 2 takes to make a complete orbit is longer than the orbital period of planet 1 by a factor 2 to the 3/2 power: that is, take 2, cube it to get 8, and take the square root of 8 to get 2.83. The year of planet 2 is 2.83 times longer than the year of planet 1.

The law would still be true, and all the behavior of the planets would be much simpler (but much less interesting), if only Plato had been right and the orbits of the planets were perfect circles. A circle can be thought of as an especially simple ellipse. Starting from an ellipse,

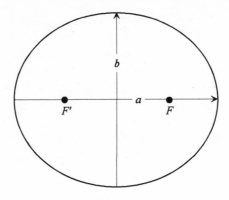

a circle can be constructed by moving both foci, F' and F, to the center:

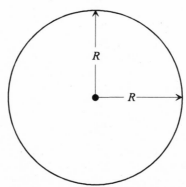

Then the semimajor axis a will be the same length as the semiminor axis b, and we will call both of them the radius, R. Notice that since a circle *is* an ellipse (a special case of an ellipse, to be sure), Kepler's laws allow planetary orbits to be circles but don't require it. In reality, the orbits of the planets in our solar system are all very nearly (but not exactly) circles—although other objects obeying Kepler's laws (such as Halley's comet, for example) have orbits that are very far from circular.

Getting back to our point, we wish to demonstrate that Kepler's third law means that the force of the Sun's gravity diminishes as the square of the distance from the Sun. Following Feynman, we'll simplify the argument by pretending that the planetary orbits really are circles. Symbolically, we'll call the time to complete an orbit T. Then Kepler's third law says $T \sim R^{3/2}$ (read, "T goes as, or is proportional to, $R^{3/2}$"), where R is the distance to the Sun. How is that related to the R^{-2} law?

Like Feynman, we are unable to follow Newton's argument here, and even Feynman's argument is a bit cryptic, so we've formulated our own. This argument is designed not only to make the point about Kepler's third law and Newton's R^{-2} law, but also to introduce some geometrical techniques we'll need for the grand finale.

The diagram that we (and Feynman) have copied from Newton shows successive *positions* of a planet in space:

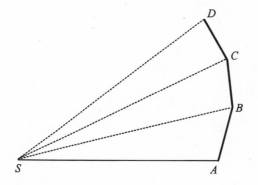

In equal intervals of time, the planet moves from A to B, from B to C, and so on. We can also represent on this diagram the velocity of the planet during each segment (due to inertia, the planet moves from A to B at constant velocity, from B to C at constant velocity, and so on). The velocity can be represented by an arrow pointing in the direction of motion (remember that the word "velocity," as it is used in physics, means not just speed but also direction):

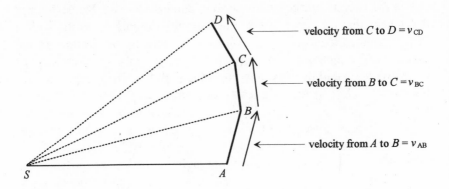

There is no reason for the velocity arrows to be drawn next to the corresponding line segment of the orbit; we can collect them together on the side at a common origin:

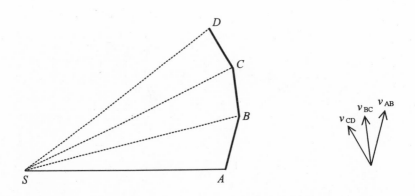

The new diagram is a velocity diagram rather than a position diagram. The direction of the arrow shows the direction of the planet's motion, so v_{AB} must be parallel to AB,

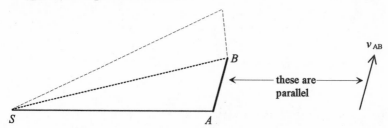

and the length of the arrow is proportional to the speed. In other words, the faster the planet is moving in that segment, the longer the arrow. If the planet moves more slowly on the segment from B to C than it did from A to B, we might get a diagram like this:

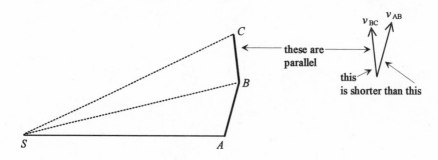

However, the change in velocity, according to Newton's second law, must be in the direction of the Sun, at point B, where the impulsive force causes the velocity to change: If v_{AB} is the velocity before the change,

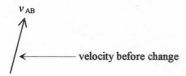

and v_{BC} is the velocity after the change,

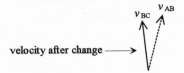

velocity after change ⟶

then the change in velocity is also an arrow,

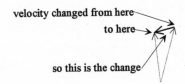

velocity changed from here
to here
so this is the change

and that arrow must be in the direction of the line from B to S:

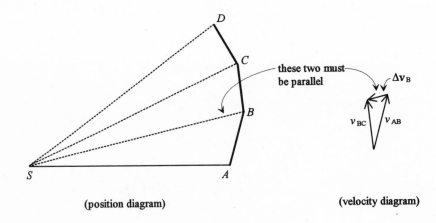

these two must be parallel

(position diagram)   (velocity diagram)

The change in velocity at point B, Δv_B, is thus in the direction of the force from the Sun, and is also proportional to the strength of the force. If the Sun's force were twice as big at point B, Δv_B would be twice as big. That's the meaning of Newton's second law. The change in velocity at each of the (imaginary) points A, B, C, . . . on Newton's diagram also depends on the (equal) time intervals between those points. Newton can (and does) imagine approximating the same orbit by time intervals half as big, to get closer to the actual smooth curve that the orbit makes in space. If all else is the same, and the time intervals are half as big, then each change in velocity will also be half as big but there will be twice as many of them:

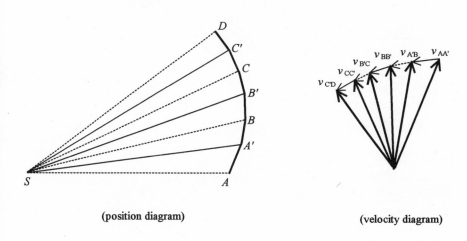

(position diagram)   (velocity diagram)

This is the same orbit, produced by the same force as the previous diagram. The force is proportional to the change in velocity at each point (half as big for this diagram) divided by the time interval (also half as big): $F \sim \Delta v / \Delta t$, where F is the force and Δt is the time interval. The force in this diagram is the same as the force in the previous diagram.

There is, as we have seen, an actual correspondence between *direction* on the position diagram and on the velocity diagram. However, the *sizes* of the diagrams bear no relation to one another at all. We could choose to make the entire velocity diagram twice as big (which wouldn't change any of the directions) and it would still be correct:

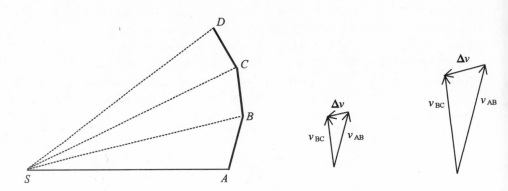

both these velocity diagrams are correct

Let's look at the simplest possible specific example. Suppose the orbit were just a circle, of radius R. Then the Newtonian diagram would look like this:

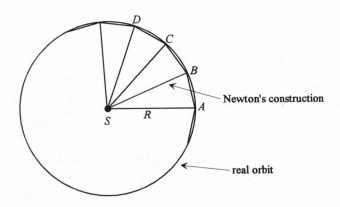

Each of the distances—*SA, SB, SC,* and so on—would be equal to *R,* the radius of the circle. Also, each change of velocity, due to the impulsive force at *A, B, C, D,* and so on, would be the same no matter how the force from the Sun depends on distance, because all these points are at the same distance from the Sun. It follows that the speeds along *AB, BC,* and so on must all be the same, and the lengths of the segments *AB, BC,* and so on are all the same. That's the only way the orbit can follow the same path, time after time. In other words, the figure drawn by Newton is a *regular polygon,* a figure of equal sides and angles, inscribed in the circle, which is the real orbit.

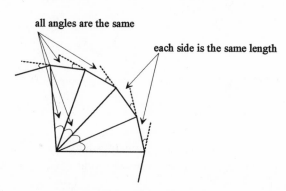

all angles are the same

each side is the same length

Regular polygons include the equilateral triangle, the square, the pentagon, the hexagon, and so on. The more sides a regular polygon has, the more it resembles a circle. Newton imagined using shorter time intervals for his figure, giving a regular polygon with more sides,

and thus more closely approximating the real circle, ad infinitum, until the real orbit is achieved.

In the velocity diagram for a circular orbit, all the velocities are of equal length and at equal angles apart, so that all the changes Δv are the same:

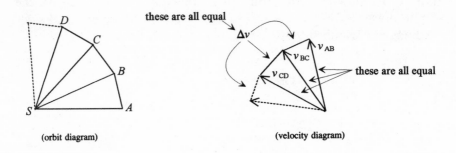

(orbit diagram)   (velocity diagram)

Thus the velocity diagram is also a regular polygon, which also becomes a circle when the orbit becomes a circle (after going through the ad-infinitum):

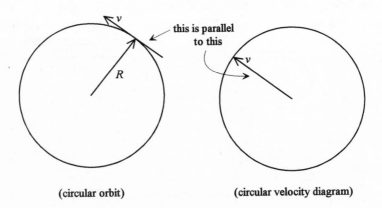

(circular orbit)   (circular velocity diagram)

The radius of the circle in the velocity diagram is v, the uniform speed of the planet all the way around its orbit. That speed is given by the distance the planet travels divided by the time it takes. The distance the planet travels is the circumference of the orbit—that is, $2\pi R$—and the time that the planet takes to go around is just what we have called T, the period of the orbit. Therefore, v is equal to $2\pi R/T$.

circular orbit:

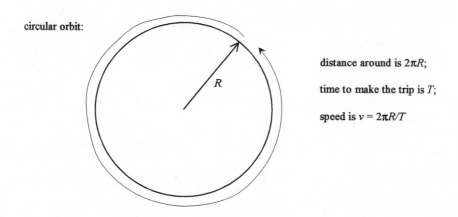

distance around is $2\pi R$;

time to make the trip is T;

speed is $v = 2\pi R/T$

Each time the planet makes one complete orbit, the velocity arrow also goes around one whole cycle:

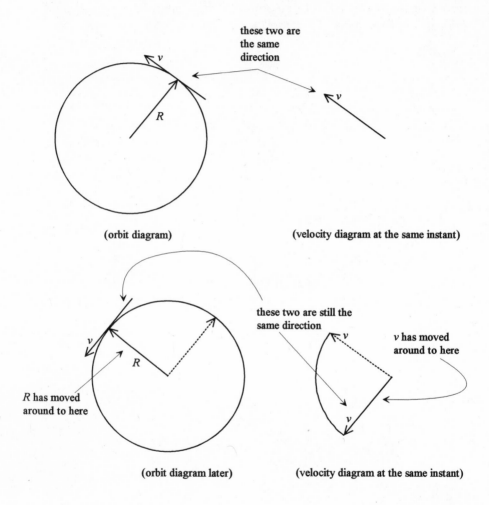

(orbit diagram)

(velocity diagram at the same instant)

(orbit diagram later)

(velocity diagram at the same instant)

When the velocity arrow makes a complete circle, the tip of the arrow moves a distance $2\pi v$:

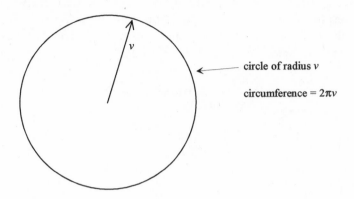

circle of radius v

circumference $= 2\pi v$

Remember that the change in velocity is given by the motion of the tip of the velocity arrow:

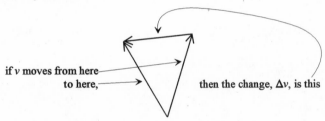

if v moves from here
to here,   then the change, Δv, is this

Let's say, now, that the circle has been divided up into 30 parts, each representing the motion in 1/30th of the orbit time T.

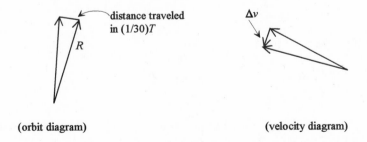

distance traveled
in $(1/30)T$

R

Δv

(orbit diagram)   (velocity diagram)

The force, as we've seen, is proportional to $\Delta v / \Delta t$, where Δv is the change in velocity, equal to 1/30th of the perimeter of the velocity circle, and Δt is the time interval, 1/30th of T. Obviously, 1/30th of the perimeter divided by 1/30th of the time is the same as the whole perimeter divided by the whole time. So $\Delta v / \Delta t$ is equal to the perimeter—that is, $2\pi v$—divided by the time T:

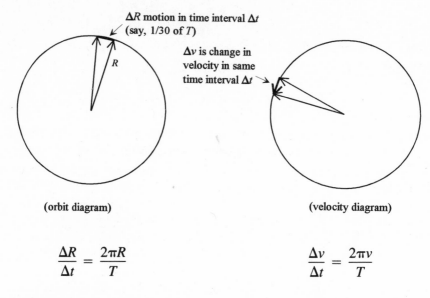

ΔR motion in time interval Δt (say, 1/30 of T)

R

Δv is change in velocity in same time interval Δt

(orbit diagram)   (velocity diagram)

$$\frac{\Delta R}{\Delta t} = \frac{2\pi R}{T} \qquad\qquad \frac{\Delta v}{\Delta t} = \frac{2\pi v}{T}$$

So the force, F, is proportional to $2\pi v/T$; and the velocity, v, is equal to $2\pi R/T$. Symbolically:

$$F \sim \frac{2\pi}{T}\, v = \frac{2\pi}{T} \left(\frac{2\pi R}{T} \right)$$

Multiplying the two fractions gives

$$F \sim (2\pi)^2\, \frac{R}{T^2}$$

This statement means, for example, that if there were a planet twice as far from the Sun (at $2R$ rather than R) and if it made its orbit in the same time period, then the force on it from the Sun, being proportional to R, would have to be twice as big. However, that's not the way planets behave. We have seen that if there were a planet at $2R$, its period would be $2.83T$. This is determined by Kepler's third law:

$$T \sim R^{3/2} \text{ (the period of a planet is proportional to the 3/2 power of its distance from the Sun)}$$

The force, F, is proportional to the distance, R, divided by T^2. But T^2 means the square of $R^{3/2}$, and $(R^{3/2})^2 = R^3$. So the force is proportional to the distance, R, divided by the cube of the distance, R^3. But R divided by R^3 is the same as 1 over R^2! The force is proportional to 1 over the square of the distance to the Sun! This is the connection we've been looking for—the R^{-2} force law.

Before plunging ahead, this is a good place to stop for a moment, to see where we've been and where we are going.

Kepler has given us three laws, and Newton has given us three laws. Kepler's laws, however, are of a vastly different character from Newton's. Kepler's laws are generalizations of observations of the heavens. They are what we would today call curve-fitting. Kepler took a few points in space—the observed positions of the planet Mars at known times—and said, "Aha! All these points fall on a curve called an ellipse!" That description trivializes the life's work of one of history's great geniuses, but it is nevertheless a correct approximation. That is the essential nature of all three of Kepler's laws.

Newton's laws are of a radically different kind. They are really assumptions about the innermost nature of physical reality: the relations between matter, forces, and motion. If the behavior deduced from those assumptions is observed in nature, then the assumptions may be correct, and if that is the case then we have seen into nature's heart, or the mind of God, depending on your taste in metaphors. In the crucially important arena of planetary motions, the test of whether the Newtonian assumptions are correct is whether they give rise to the Keplerian laws, which summarize with great precision an immense amount of astronomical data.

The connection between Newton's laws and Kepler's laws is more complex than that, however; so far, there is a missing link. In order to determine the planetary motions that his laws would dictate, Newton had to discover the nature of a particular kind of force—the force of gravity. In order to do so, he made use of Kepler's second and third laws. Then, having thus deduced the nature of gravity, he was able to demonstrate that the force of gravity, acting under the direction of his laws, would produce Kepler's remaining observation, the law of ellipses. That is the logical sequence of events as presented by Newton in his *Principia*. We now stand at the point in his argument where we have deduced the nature of gravity, making use of Newton's laws and Kepler's second and third laws. Let's review how we did that, before the curtain rises on our final act—Kepler's first law, the law of ellipses.

As applied to planetary motions, Newton's first law, the law of inertia, says that if a planet has no force acting on it, it will remain at rest if it begins at rest, or it will move forever in a straight line at constant speed if it begins in motion. Why it does so is a mystery, although Newton sometimes refers to the mechanism as the planet's "inner force." However, the point with regard to Newton's laws is not to ask *why* they are true, but to ask only *whether* they are true.

Newton's second law says that if there is indeed a force F acting on a planet, its effect is to divert the planet from the straight line that the planet would have followed at constant speed under the influence of inertia. In particular, if a force is applied for a given time interval, Δt, it produces a change in velocity—that is, a departure from the inertial path, Δv, proportional to the force and in the same direction as the force. That means that if twice the force $(2F)$ is applied, then twice the change in velocity $(2\Delta v)$ is produced. It also means that $2\Delta v$ can be obtained by applying the same force for twice the amount of time $(2\Delta t)$. Symbolically, we would write $\Delta v \sim F\Delta t$. It further means that if the force is toward the Sun, the change in velocity must be toward the Sun.

Newton's third law says that forces which operate between different parts of a planet produce no net force upon the whole planet, so that, for purposes of analyzing planetary motions, we can ignore the fact that planets are large complicated bodies and treat them as if they were concentrated at a mathematical point at their centers.

The picture Newton then pursues is that the Sun, assumed to be immovable, applies on the planets a force, gravity, that diverts them from the inertial straight lines they would otherwise follow and into their actual orbits.

One property of those actual orbits, described by Kepler's second law, is that a hypothetical line connecting the Sun to a planet sweeps out equal areas in equal times as the planet moves around in its orbit. Newton shows, and we have now shown, that the meaning of Kepler's observation is that the force of gravity acts in the direction of the line connecting the planet to the Sun.

A second property of planetary motion is that the farther away a planet's orbit is from the Sun, the more slowly the planet moves in that orbit. Specifically, the time the planet takes to make one complete circuit increases as the 3/2 power of the distance of its orbit from the Sun. Newton shows, and we have now shown, that to produce this result, the force deflecting the planets into their various orbits must weaken as 1 over the square of the distance from the Sun. In other words, if a planet is twice as far from the Sun, the gravitational force attracting it toward the Sun will be four times smaller.

Notice that Kepler's second law (equal areas) deals with the motion of a single planet in different parts of its orbit, while his third law compares the orbits of different planets. It is strange but true that the masses of the planets have no bearing at all on how fast they move in their orbits. A year (one complete orbit) of the planet Earth is shorter than a year of the planet Jupiter only by the ratio of the 3/2 powers of their distances from the Sun, though Jupiter's mass is more than 300 times that of the Earth.

In any case, we now know that the force of the Sun's gravity on a planet is directed toward the Sun, and that its strength decreases as 1 over the square of the distance from the Sun. We have used Kepler's second and third laws to find out that much. The final, triumphant accomplishment will be to show that such a force of gravity, acting as directed by Newton's laws, will produce elliptical orbits for the planets.

In Feynman's lecture, this is the point at which he finds himself unable to follow Newton's line of argument any further, and so sets out to invent one of his own. His first departure from Newton is much like some brilliant, completely unexpected move by a chess prodigy. Instead of dividing the orbit into imaginary segments that take equal intervals of time, as Newton always does, Feynman divides the orbit into segments that make equal angles at the Sun. We'll need to sketch some diagrams to see what this means.

Recall the diagram in the *Principia* that Feynman copied into his lecture notes:

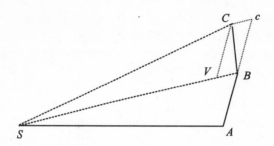

In a certain time interval, a planet would move from A to B if there were no force from the Sun. The time interval might be, for example, 1 second, or 1 minute, or 1 month. In the next equal time interval, it would continue an equal distance from B to c. Instead, the force from the Sun produces an impulse at B that dictates a change in motion, directed toward the Sun, equal to BV. During the second time interval, the planet actually executes a combination of the path Bc, dictated by inertia, and the path BV, dictated by the Sun's gravity: it follows the diagonal of the parallelogram formed by the two motions and arrives at C. We proved earlier that the triangles swept out in equal times, SAB and SBC, have equal areas. Thus Newton approximates the orbit as a series of points equally spaced in time (A, B, C, . . .) at each of which the planet is diverted from its inertial straight line by an instantaneous pull from the Sun. The shorter the time intervals, the more frequent the pull from the Sun and the more nearly the trajectory comes to resemble the real orbit, which is a smooth curve with the Sun's gravity acting continuously to pull the planet away from the inertial straight line it would otherwise have followed. The final, smooth orbit retains the property we (and Newton and Feynman) have demonstrated for the schematic one: it sweeps out equal areas in equal times, which means that the planet moves faster in its orbit when it is closer to the Sun.

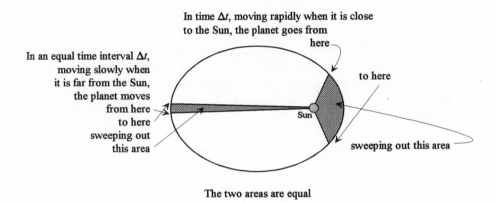

The two areas are equal

Feynman has used the same argument, taken directly from Newton, to prove this law of equal areas. Now, however, he chooses to divide the orbit into equal angles rather than equal areas:

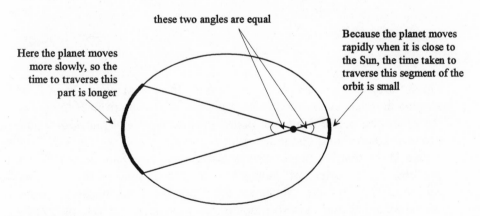

The two segments of orbit shown above have equal central angles, but they sweep out different areas and therefore take different amounts of time. The law says that the planet sweeps out equal areas in equal times. That means that if it sweeps out half as much area, it takes half as much time, or

$$\Delta t \sim (\text{area swept out})$$

Let us for the moment represent these equal-angle segments on a Newton-type diagram, on which the planet undergoes inertial straight-line motions punctuated by velocity changes due to the force of gravity. For simplicity, we draw the velocity changes, Δv, directly on the orbit diagram:

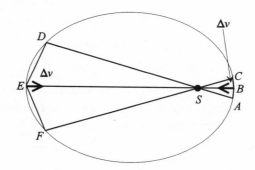

On the side of the orbit closer to the Sun, the planet glides from A to B, gets diverted by Δv due to the Sun, and continues from B to C. On the other end of the orbit, the planet goes from D to E, suffers a pull producing a Δv, and continues from E to F.

We know that the planet moves faster along BC than along EF. To see how much faster, we have to compare the areas of the triangles SBC and SEF, because the times are proportional to the areas swept out. Remember that the two triangles have the same central angle at S. Reorienting SEF and laying it on top of SBC, we have:

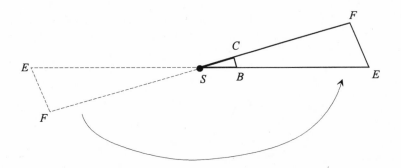

The area of each triangle is 1/2 (the base) × (the altitude). Also, these are similar triangles. That means that if the base of the larger triangle is twice as big as the base of the smaller one, then the altitude is also twice as big; in that case, the area of the big triangle would exceed the area of the small one by 2 × 2 = 4. The general rule is that the area is proportional to the square of the distance from the Sun.[2] So, the time it takes to go through any portion of the orbit is proportional to the area swept out, which is proportional to the square of the distance from the Sun. Here's a comparison of Newton's way and Feynman's way of dividing the orbit into segments:

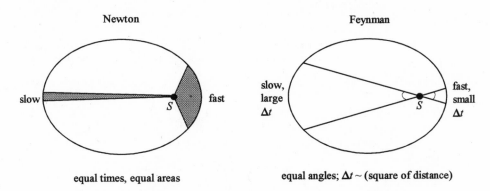

[2] In his lecture, Feynman glosses over this point in a single line. It is not so simple, however, and we haven't really proved it either. Here's a more complete proof. Consider two arbitrary orbit segments that have equal central angles:

(footnote continued)

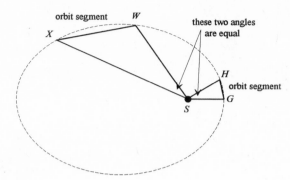

Lay the triangle *SWX* on top of *SGH* like this:

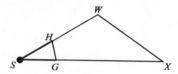

It is always possible to draw a line through *WX*, parallel to *HG*, such that the two little triangles that result will have equal areas:

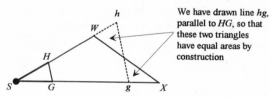

We have drawn line *hg*, parallel to *HG*, so that these two triangles have equal areas by construction

The triangle *Sgh* has the same area as *SWX* (it is bigger by one of the little triangles and an equal amount smaller by the other), and it is similar to *SGH*. Now draw a line from *S* to the point where *WX* crosses *hg*:

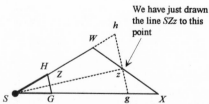

We have just drawn the line *SZz* to this point

We will now call *SZ*, or *Sz*, the distance from the Sun to the orbit. According to the property of similar triangles (base and altitude each increase as the size, so the area is proportional to the square of the size), the similar triangles *SGH* and *Sgh* have areas in proportion to the squares of the lengths *SZ* and *Sz*. But *SWX* has the same area as *Sgh*, so the area of *SWX* is also in proportion

Symbolically, $\Delta t \sim R^2$ in the Feynman drawing, where R is the distance from the planet to the Sun. But we also know that the force from the Sun decreases with distance, according to the inverse-square law—that is, $F \sim 1/R^2$. Let's go back to the kind of diagram that shows the change in velocity, Δv, at each discrete point of the orbit:

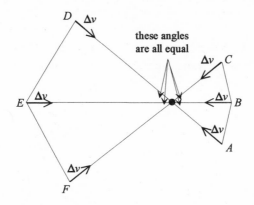

At each point around the orbit—A, B, C . . . D, E, F . . . , and all the points in between—there is a Δv toward the Sun. The bigger the force F, the bigger the Δv; also, the longer the time interval Δt, the greater the change in velocity Δv:

$$\Delta v \sim F \, \Delta t$$

But since $F \sim 1/R^2$ and $\Delta t \sim R^2$,

$$\Delta v \sim (1/R^2) \times R^2 = 1$$

This means that Δv does not depend on R at all! Everywhere in the orbit, no matter how close to the Sun or how far away, the Δv produced in a given *angle* is the same. That happens, as we have now seen,

to the square of Sz. If we now imagine shrinking the central angle down smaller and smaller ad infinitum, the line SZz always stays inside the angle, and because the points W and X on the elliptical orbit get closer and closer together, the length Sz ultimately becomes equal to SW or SX, which is what we previously called the distance to the Sun. QED.

because as the planet gets farther away from the Sun, the force acting on it gets weaker (as the square of the distance) but the time the force has to act on the planet gets longer (also as the square of the distance). The result is that all the Δv's, all the way around the orbit, are the same. That, says Feynman in his lecture, is "the central core from which all will be deduced—that equal changes in velocity occur when the orbit is moving through equal angles."

To see exactly what this means, let us look back for a moment at the type of diagram sketched by Newton and copied by Feynman. Rather than representing positions of the planets, we will represent velocities:

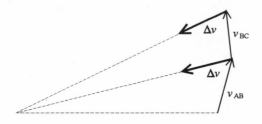

In Newton's way of doing things, the time intervals were all the same, and the Δv's were all pointed toward the Sun, but some Δv's were bigger than others (the biggest Δv's came when the planet was closest to the Sun). In Feynman's scheme, the central angles are all the same, so that the time intervals are different. The Δv's all point toward the Sun (they must, according to Newton's second law) and *they are all now exactly equal in size, all the way around the orbit*. This has consequences that are now to be worked out.

At this point, Feynman has sketched in his lecture notes, with meticulous care, the orbit diagram and the corresponding velocity diagram for equal-angle segments. Here is the result:

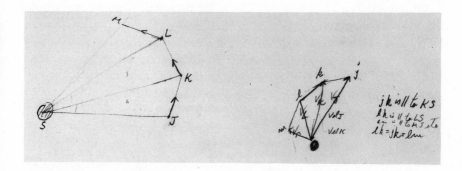

The orbit starts from position J, goes to K making some angle at the Sun, suffers a Δv changing its direction, then continues through an equal angle from K to L, and then again from L to M:

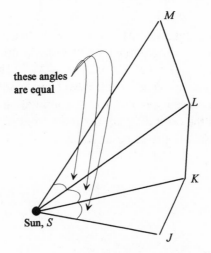

Unlike Newton's version of this diagram, the times of these segments are not necessarily equal. The velocities are in the directions JK, KL, and so on. They are, in general, of different magnitudes on different segments. The changes in velocity suffered at points J, K, L, and M are all directed toward the Sun and all of the same magnitude. In other words, at J there is a Δv in the direction JS; at K, the same Δv occurs in the direction KS; and so on. Using these facts, Feynman constructs the velocity diagram:

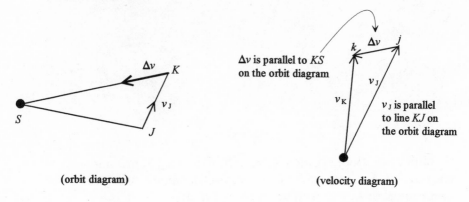

(orbit diagram)   (velocity diagram)

On the orbit diagram, the planet moves from J to K with velocity v_J. On the velocity diagram, v_J has the same direction, but not the same length, as JK. At point K, there is a Δv in the direction KS, moving the velocity diagram a distance Δv from point j to point k, where the velocity becomes v_K. This process continues at the next step; the second segment on the orbit diagram is drawn from K, parallel to v_K, to a point L, so that KSL is the same angle as JSK:

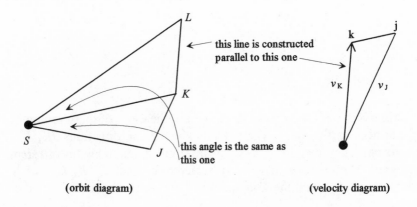

(orbit diagram)   (velocity diagram)

We now find the point *l* on the velocity diagram by adding a Δ*v* equal in magnitude to *jk*, but parallel to *LS*:

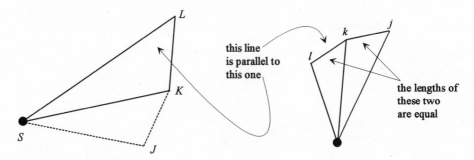

The same procedure can be repeated all the way around the orbit. The next step gives the diagram as Feynman sketched it in his notes:

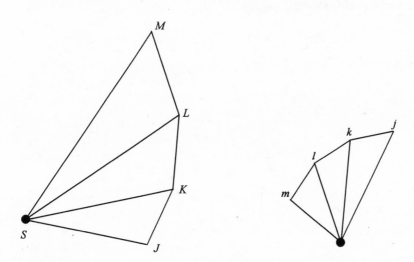

As Feynman wrote in his notes, *jk* is parallel to *KS*, *lk* is parallel to *LS*, *lm* is parallel to *MS*, and *lk* = *jk* = *lm*.

Each of the sides of the velocity diagram (*jk*, *kl*, *lm*, . . .) is parallel to one of the lines radiating from the Sun in the orbit diagram. Because the lines from the Sun are constructed to have equal angles, the sides of the figure in the velocity diagram also have equal external angles:

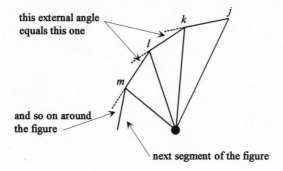

this external angle equals this one

and so on around the figure

next segment of the figure

When the velocity diagram is complete, it will be a figure with equal sides and equal (external) angles:

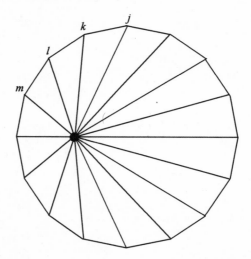

Notice that the velocities themselves, which are the distances from the origin to j, k, l, and so on, are unequal but that the sides (the Δv's) are equal. The resulting figure is a regular polygon! The origin of the velocities is not at the center, but the external figure itself is a regular polygon.

If we now proceed as usual to divide the orbit diagram into a larger number of segments with equal but smaller angles, the orbit more nearly approaches a smooth curve—and so does the velocity diagram. Because the velocity diagram is a regular polygon, the smooth curve it approaches is a circle! But the origin of the velocities is not necessarily at the center of the circle.

At this point, Feynman sketches in his lecture notes the orbit and velocity diagrams as smooth curves. First the orbit. It starts at point J, and Feynman has drawn it in the conventional way, with the line from the Sun extending horizontally; in contrast to the segmented orbit diagram, the velocity at point J is a vertical line, perpendicular to the line from the Sun:

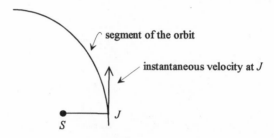

After some time, the planet arrives at point P, having made an angle θ at the Sun:

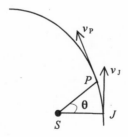

At each point, the instantaneous velocity is tangent to the smooth curve.

Now construct the corresponding velocity diagram. It will be a circle, with the origin off-center. The length of the line we will draw to represent v_J will depend on the planet's speed at point J of the orbit. Remember that on a velocity diagram, the longer the line, the faster the speed. Point J on Feynman's orbit diagram is also the closest point to the Sun (Feynman has decided this in his head without mentioning it in the lecture), where the orbital speed is greatest. Therefore the line v_J must pass through the center of the circle, because it has to be the longest line on the velocity diagram:

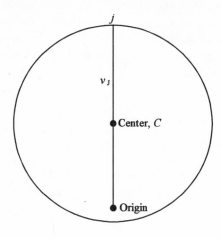

Drawn this way, v_J is vertical (parallel to v_J on the orbit diagram), and it is the longest distance from the origin to any point on the circle. The velocity at point p on the velocity diagram, corresponding to P on the orbit diagram, is a line from the origin parallel to v_P:

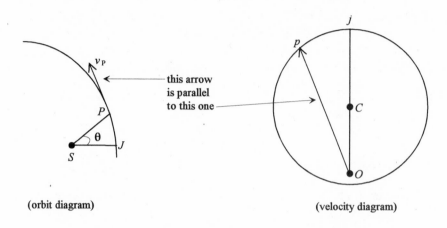

(orbit diagram)   (velocity diagram)

It is also true that the angle jCp on the velocity diagram is the same angle, θ, as JSP on the orbit diagram:

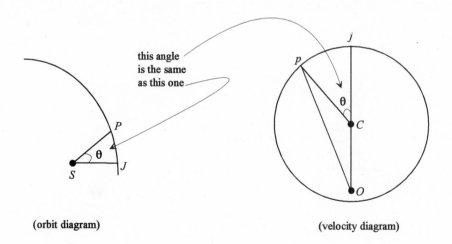

(orbit diagram)   (velocity diagram)

The reason for this can be seen if we go back to the complete velocity diagram of orbit segments—the regular polygon—and draw lines out from its center instead of from the origin of the velocity arrows:

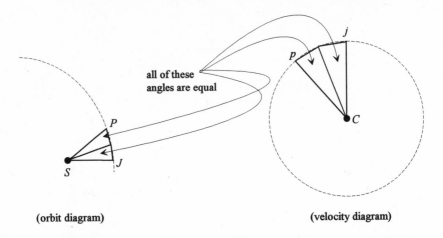

all of these
angles are equal

(orbit diagram)   (velocity diagram)

The orbit has been divided up into some number of equal angles, which must total 360°. The polygon necessarily has the same number of equal sides, each occupying the same fraction of 360°. Therefore the angle from SJ to any point on the orbit is the same as the angle from Cj to the corresponding point on the velocity diagram.

The net result is shown in the pair of diagrams sketched by Feynman:

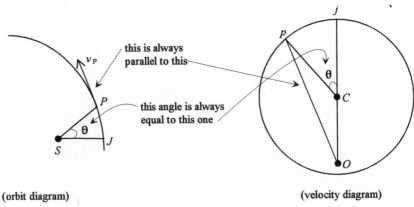

this is always
parallel to this

this angle is always
equal to this one

(orbit diagram)   (velocity diagram)

Now that all the correspondences between the two diagrams have been established, we could construct the orbit starting from the velocity diagram. It is an easier starting point, because we know that it is just a circle:

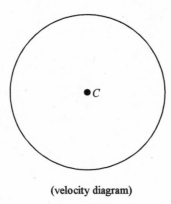

(velocity diagram)

Any orbit permitted by Newton's laws and the force of gravity will have this same velocity diagram. The exact shape of the orbit will depend on where we choose to place the origin of the velocities. Pick a point, any point, inside the circle, but not at C, the center (we will see later what happens if the point *is* at C, or on the circle, or even outside it):

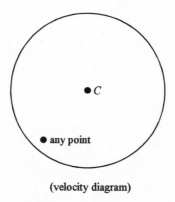

(velocity diagram)

For purposes of familiarity only, turn the whole diagram until the chosen point lies directly below C:

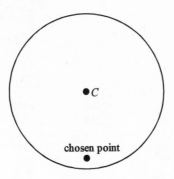

The chosen point is to serve as the origin of velocities: that is, a line from there to any point on the circle's perimeter will have a length proportional to the planet's speed at that point on the orbit, and lie in the same direction as the planet's motion at that point on the orbit. As noted, the line from the origin through the center to the circle's perimeter is the longest line and therefore represents the point on the orbit where the planet is moving fastest.

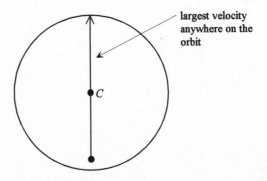

According to the equal-areas law, this will be the point on the orbit closest to the Sun. As Feynman has done, we will draw the orbit so that the line from there to the Sun is horizontal and the velocity is vertical (that's why we rotated the origin of the velocity diagram to be beneath the center):

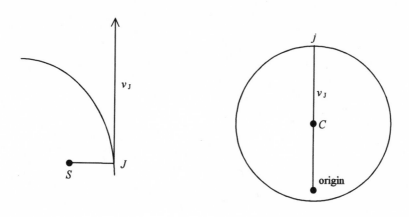

Now draw a line from the origin to any other point on the circle, p:

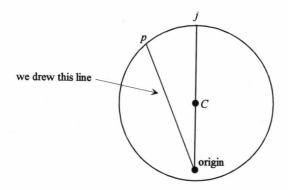

This point corresponds to a point P on the orbit that has the following properties: the line from the origin to p on the velocity diagram is parallel to the tangent at the point P on the orbit diagram, and the angle jCp is the same as the angle JSP:

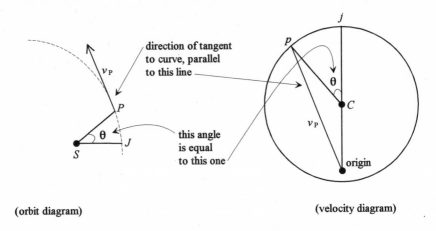

(orbit diagram)   (velocity diagram)

So at each angle θ, we know the direction of the tangent to the orbit we are seeking to construct. How can we construct the curve?

Later in the lecture, Feynman tells us that this was the most difficult step to discover. The trick is to rotate the velocity diagram clockwise by 90°, so that the directions on it are the same as those on the orbit diagram:

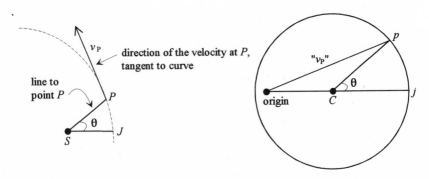

Now the central angle θ is the same on both diagrams, but the line marked "v_p," which was parallel to the velocity at P on the orbit, is now perpendicular to it, since we rotated the whole velocity diagram by 90°. We now know, from the velocity diagram, the direction from the Sun to point P on the orbit, and we know the direction of the tangent to the orbit at that point. It is perpendicular to the line marked "v_P." But we don't yet know exactly where the point is.

The easiest way to construct the curve having all the required properties is to draw it right on top of the velocity diagram. Then the size of the orbit will be arbitrary, but all the directions, and therefore the shape of the orbit, will be correct. To get the orbit, simply construct the perpendicular bisector of the line from the origin to p:

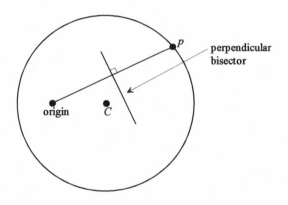

Because it is perpendicular to the line from the origin to p, we know that it is parallel to v_P, the velocity at point P on the orbit. At some point, the perpendicular bisector crosses the line connecting p to the center, C:

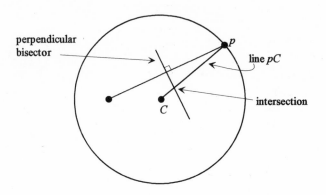

As the point p moves around the circle, the intersection of pC and the perpendicular bisector moves around in a curve of its own:

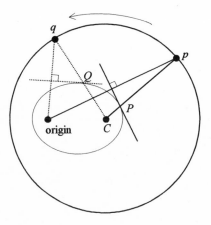

As p moves around the circle to q, the intersection of the construction moves from P to Q and so on, creating the orbit.

We once before made exactly the same construction. Starting from two points in the plane called F' and F (corresponding respectively to origin and C), we drew a line from F' to a point G' (p in the new diagram):

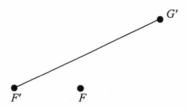

Then we connected $G'F$, and drew the perpendicular bisector of $F'G'$, which crosses FG' at the point P:

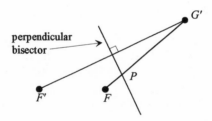

We proved then that as the point G' executes a circle centered at F, the point P executes an ellipse, and at each point P the perpendicular bisector is tangent to the ellipse (see pages 73 to 80).

We have now made exactly the same construction again as on page 79—only the names have been changed. Here's how the new diagram looks:

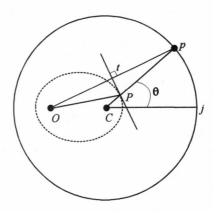

Here, p is a point on a circle centered at C. There is also an eccentric point: the origin of the velocity diagram, which we now call O. The line segment Op has a perpendicular bisector at t, which intersects the line Cp at a point P. We will now prove again that each point P created in this way, as p moves around the circle, lies on an ellipse, and that the line tP is tangent to the ellipse at P. Since tP is parallel to the velocity of the planet when it is at point P on its orbit, we will have constructed the unique curve that has the planet going in the right direction at every point in its orbit.

To prove that the curve is an ellipse, we notice that the triangles OtP and ptP are congruent:

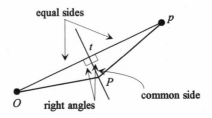

equal sides

right angles

common side

Therefore $OP = pP$. And in the full diagram,

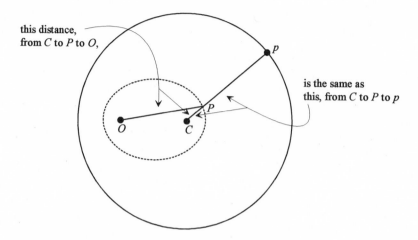

this distance, from C to P to O,

is the same as this, from C to P to p

CPp, which is the radius of the circle and is therefore the same all the way around, is equal to $CP + PO$, the length of the string from foci C and O that constructs the ellipse. The dashed curve (the orbit) is therefore an ellipse, QED. To prove that tP is the tangent line at P, go back to the congruent triangles:

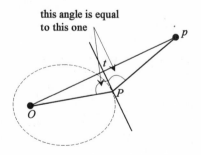

this angle is equal to this one

Now let the lines *Pp* and *tP* cross each other:

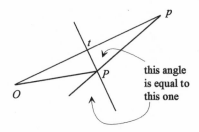

this angle
is equal to
this one

Therefore,

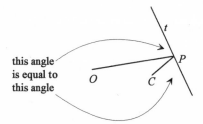

this angle
is equal to
this angle

The line *tP* is therefore the line that reflects light from *C* to *O* at point *P*. We long ago proved that the line *tP* that has that property is the tangent line. For the last time, QED.

The proof is now complete. Feynman is not quite finished yet, but we have accomplished in full what we set out to show. Newton's laws, together with an R^{-2} force of gravity toward the Sun, result in elliptical orbits for the planets. Before we leave the subject, let us look back one more time at the logic of the arguments that have enabled us (with the help of Newton and Feynman) to accomplish that heroic feat.

Newton says something like this: From the fact that planets sweep out equal areas in equal times, I used my laws to deduce that the force of the Sun's gravity on a planet points directly toward the Sun. Then, from the fact that the orbital periods of planets are proportional to the 3/2 power of their distances from the Sun, I used my laws to deduce that the force of gravity diminishes as R^{-2}. Finally, my laws, together with these two facts about gravity, produce elliptical orbits.

Newton didn't really think about the problem that way. We know from earlier versions of his work (for example, the brief treatise he sent to Halley in 1684) that he experimented with various forms of his axioms about dynamics. Only later did he reduce them to three and start to refer to them as "laws." The act of reducing all of dynamics to three fundamental laws was supremely important, because, as Newton and his followers were to show over the course of the ensuing three centuries, those laws could be used to explain not only the motions of the planets but almost every other phenomenon in the physical world as well. Newton's laws tell us how matter behaves when it is acted on by forces. The only two things we need to know about the physical world that Newton's laws don't tell us are: What is the nature of matter? What is the nature of the forces that act between bits of matter? These two questions are still the central concerns of the science of physics.

This whole powerful reorganization of our understanding of the world begins with the proof of elliptical orbits. In this case, we do not need to know very much about the nature of matter, because gravity affects all matter in exactly the same way. The nature of the force of gravity is very important, however, and that's what Newton uses two of Kepler's laws to deduce.

Finally, we have seen the proof of elliptical orbits not as Newton originally did it but as Richard Feynman worked it out. Feynman divides the orbit into equal angles. In each equal-angle segment, the change in velocity is directed at the Sun, and proportional to the strength of the force and the time over which the force acts. That is Newton's second law. The time is proportional to the area swept out, which (by pure geometry) is proportional to the square of the distance, and the force is inversely proportional to the square of the distance (that's the nature of the force of gravity); so no matter what the shape of the orbit is, and no matter how close to or far from the Sun the planet wanders, the planet undergoes equal changes of velocity in equal angles. It follows immediately that the velocity diagram is a regular polygon (equal sides at equal angles), which becomes a circle for smooth orbits. However, the origin of the velocity diagram *is not at the center of the circle*.

Then, with the help of a geometric construction that has been cunningly set up in advance, it is shown that the orbit has the shape of an ellipse, with the origin of the velocity diagram and the center of the velocity circle acting as foci.

The velocity diagram is a powerful geometric tool. Newton's dynamical laws, together with an R^{-2} force, always produce a circular velocity diagram:

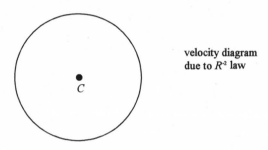

velocity diagram
due to R^{-2} law

The shape of the orbit depends on where O, the origin of the velocity diagram, is. If O coincides with C, the center of the diagram, then the two foci of the ellipse coincide and the planet has the same speed in all parts of its orbit:

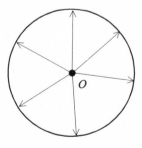

In this case, the orbit is simply a circle.

If the point O is anywhere between C and the circumference of the diagram, then the orbit is an ellipse. The closer O is to C, the more nearly circular is the ellipse. The farther O is from C, the more elongated the ellipse:

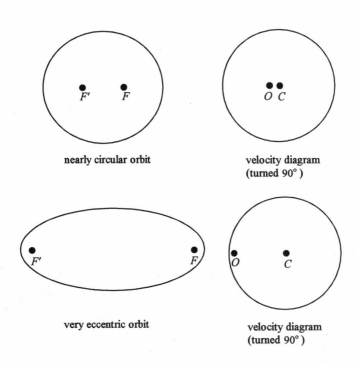

nearly circular orbit

velocity diagram
(turned 90°)

very eccentric orbit

velocity diagram
(turned 90°)

In our solar system, all the planetary orbits are nearly circular. In the Earth's orbit, the distance between foci is about 1 percent of the diameter of the orbit; for Mars, it is about 9 percent; for Mercury and Pluto (whose orbits are the most eccentric), a little more than 20 percent. Halley's comet, by contrast, has an extremely eccentric elliptical orbit. The distance between its foci is 97 percent of the diameter of its orbit.

What happens if O is *outside* the circle? Let's go back to the velocity diagram before we turned it by 90°. We still have the largest velocity in the orbit at the point of closest approach:

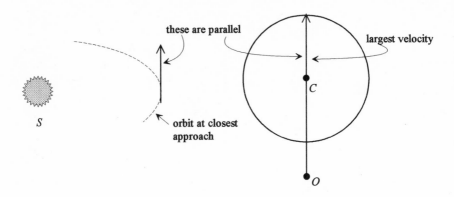

As the angle θ increases, the velocities proceed around the circle in the diagram:

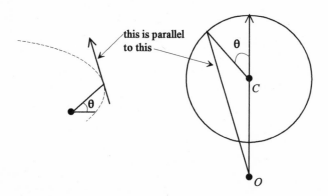

At some value of θ, the line from O is the tangent to the velocity circle:

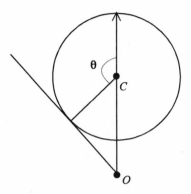

Remember, this line is also parallel to the instantaneous velocity of the orbit and the tangent to the velocity diagram is in the direction of the Δv's in the orbit diagram, which represent the changes in the velocity. In other words, at this angle θ, the change in velocity is in the same direction as the velocity itself. That means the velocity is *not* changing direction anymore. The path is no longer a curve, it is a straight line. The "orbit" is therefore not an ellipse, on which the path is never a straight line. Instead, it is a hyperbola, another of the conic sections, which tends to become a straight line far away from the focus:

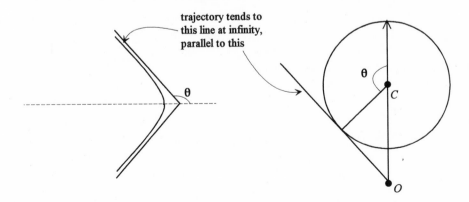

trajectory tends to this line at infinity, parallel to this

On this trajectory, the "planet" falls toward the Sun from infinity, swings around, and escapes back to infinity. Its path is not an orbit at all. When it starts from infinity, and when it gets back there, its velocity is not zero; the velocity is proportional to the length of the line from O to the point where it is tangent to the velocity circle.

If the point O is *on* the circle, the "planet" also escapes to infinity, but it has zero velocity when it gets there; this trajectory is a parabola. Thus, Newton's dynamics together with an inverse-square force give circular velocity diagrams. Depending on where the origin of the velocity diagram is, the orbit can be a circle, an ellipse, a parabola, or a hyperbola—the curves collectively known as the conic sections.

In the very last part of his lecture (just because he has time left over, he says), Feynman turns the machinery he's developed onto a very different kind of problem—and again, one of vast historical significance.

In 1910, two researchers, Ernest Marsden and Hans Geiger, acting at the suggestion of their leader, Ernest Rutherford, found that if a beam of α (alpha) particles (the nuclei of helium atoms) was directed at a thin gold foil, a few of them would be scattered backward instead of passing through the foil. The experiment might be thought of as crudely analogous to some alien being firing a comet into the solar system in an attempt to determine whether the mass of the solar system was spread out in a uniform blob or mostly concentrated in a compact object (the Sun) at the center. Only a compact object could have any hope of turning the comet around and hurling it back. Instead of a comet, Rutherford's group had the α particle, and instead of the solar system, atoms of gold. The question was whether the matter inside an atom was spread out more or less uniformly (as current theory then held) or was concentrated at the center. The fact that some α particles were scattered backward showed that the mass had to be concentrated at the center, and this experiment constituted the discovery of the atomic nucleus.

Here, the force operating between the projectile and the constituents of the system was not gravity but electricity. Electricity is a force that acts between positive and negative electric charges (terms coined by a self-educated Newtonian scientist of the eighteenth century, Benjamin Franklin). Like gravity, electricity is an R^{-2} force that acts along the line joining the charges; unlike gravity, it can either attract charges toward each other (opposite charges) or cause charges to repel each other (like charges). The force of gravity always attracts, never repels. The electric force is vastly more powerful than the gravitational force. In fact, it is so powerful that it is self-neutralizing. Every atom in the gold foil has exactly the same amount of positive and negative charge, so from the outside the atom is neutral, exerting no electric force if it is not disturbed. The question is: What happens when an electrically charged projectile—the α particle, which is electrically positive—is fired into an atom? The answer is that it is repelled by the atomic nucleus, which contains all the positive charge and nearly all the mass of the atom. Occasionally, by sheer chance, an α particle will come close enough to the nucleus to get kicked almost directly backward. That's what Marsden and Geiger observed.

Because electricity is an R^{-2} force acting along the line between the charges, then if the particles obey Newtonian dynamics all the geometric arguments that Feynman used earlier are applicable to this problem. This problem is to find the probability that a projectile will be kicked back, so that the experiment can be compared to a quantitative theory. The starting point is the velocity-diagram circle (good for any R^{-2} force along the line between the particles), with the origin outside the circle. The "orbits" of the α particles will not be ellipses trapped forever in the vicinity of the nucleus, but rather hyperbolas, which will send the α particles away to infinity after bending their trajectories through some larger or smaller angle. We will not try to follow all the steps now, because Feynman no longer feels constrained to stick to geometrical arguments. Instead he pulls out all the analytic stops in order to arrive at what is, as he says, a very famous formula.

It deserves its fame, because it led directly to the discovery of quantum mechanics, and hence to the overthrow of the Newtonian dynamics used to arrive at the formula! But that's a story for another book. Now the time has come to put ourselves directly in the hands of the master. Enter Feynman.

4

"The Motion of Planets Around the Sun"

(MARCH 13, 1964)

[Note: We advise that this chapter be read while listening to the recording of Professor Feynman's lecture.]

The title of this lecture is "The Motion of Planets Around the Sun."
. . . After the bad news you just heard announced, I have some good news for the same reason, that since the exam is coming up Tuesday, nobody wants to give a lecture that you have to study, so I'm giving a lecture that's just for the fun of it, for your entertainment [applause]. All right, all right, I won't be able to give it. Save all that for the end and then make up your mind.

The history of our subject of physics [arrived] at one of the most dramatic moments when Newton suddenly understood so much from so little. And the history of this discovery is of course the long story about Copernicus, Tycho [Brahe] making his measurements of the positions of the planets, and Kepler finding the laws which empirically describe the motion of these planets. It was then that Newton discovered that he could understand the motion of the planets by stating another law. And you know all this from the lecture on gravitation, so I continue directly from there with a quick summary of that material.

In the first place, Kepler observed that the planets went in ellipses around the Sun, with the Sun as the focus of the ellipse. He also observed—he had three observations to describe the [orbits]—that the area that's swept out by a line drawn from the Sun to the orbit is proportional, this area here, is proportional to the time. Finally, to connect planets in different orbits, he discovered that the planets with different orbits have periods, or times of rotation around the complete orbit, which bear a 3/2 power ratio to the major axis of the ellipse. If there were circles (to make it easy), it would mean that the square of the time to go around the circle is proportional to the cube of the radius of the circle.

Now, Newton was able to discover two things from this. First he noticed that equal areas and equal times meant, from his point of view about inertia, that the material would continue in a straight line at a uniform velocity if it were not disturbed, that the deviations from the uniform velocity are always directed toward the Sun, and that equal areas and equal times is equivalent to the statement that the forces are toward the Sun. So he used one of Kepler's laws already to deduce that the forces were toward the Sun. And then it is easy to argue—especially for the special case of circles from the third law—that for such circles the force which would be directed toward the Sun would have to go inversely as the square of the distance.

The reason for that is something like this. Suppose that we take a certain fractional part of an orbit, some fixed angle, a small angle, and a particle has a certain velocity in one part of the orbit and another velocity later on. Then the changes in velocity for a fixed angle are evidently proportional to the velocity. And the change in velocity during an interval of time—during a fixed time—which is the force, is evidently proportional to the velocity in the orbit times the time that it takes to go across this fraction of the orbit. I mean, divided by the time. So the velocity changes proportional to the velocity. And the time over which that change has taken place is proportional to the time that it takes to go around the whole orbit—because it is a fixed angle, like one-hundredth of the orbit. Therefore the centripetal acceleration, or change per second of the velocity in the direction of the center, is proportional to the velocity on the orbit divided by the time that it takes to go around.[1]

You can put that in many different ways, because of course the time it takes to go around is related to the velocity by this relation. That the speed times the time is the distance around—or, rather, that the speed times the time is proportional to the radius. And so you can either substitute for the time, obtaining your famous v^2/R. Or better, I'll substitute for the velocity R/T. The velocity is evidently proportional to the radius divided by the time that it takes to go around, so that the centrifugal acceleration goes as the radius and inversely as the square of the time to go around. But Kepler tells us that the time to go around squared is proportional to the cube of the radius. That is, the denominator is proportional to the cube of the radius, and therefore the acceleration toward the center is inversely as the square of the distance. So Newton was able to deduce—in fact, [Robert] Hooke deduced earlier than Newton in the same way—that this force would be inversely as the square of the distance. So from two of Kepler's laws, we come [away] with only two conclusions. No one can verify anything that way. This may be of no particular interest, because the number of hypotheses entered is equal to the number of facts checked as the number of guesses used.

On the other hand, what Newton discovered—and which was the most dramatic of his discoveries—was that the third law [Feynman means the First Law] of Kepler was now a consequence of the other two. Given that the force is toward the Sun, and given that the force varies inversely as the square of the distance, to calculate that subtle combination of variations and velocity to determine the shape of the orbit and to discover that it is an ellipse is Newton's contribution, and therefore he felt that the science was moving forward, because he could understand three things in terms of two.

As you well know, he understood ultimately many more than three things—that the orbits in fact are not ellipses, that they perturb each other, that the motion of the Jupiter satellites is also understood, the motion of the Moon around the Earth and so on, but let us just concen-

---

[1]Feynman is saying $\Delta v/\Delta t$ is proportional to v/T. See Chapter 3, page 108. He refers to $\Delta v/\Delta t$ as "the centripetal acceleration" above, and below he calls it "the centrifugal acceleration."

trate on this one item, in which we disregard the interactions of one planet with another.

I can summarize what Newton said and in this way about a planet: that the changes in the velocity in equal times are directed toward the Sun, and in size they are inversely as the square of the distance. It is now our problem to demonstrate—and it is the purpose of this lecture mainly to demonstrate—that therefore the orbit is an ellipse.

It is not difficult, when one knows the calculus, and to write the differential equations and to solve them, to show that it's an ellipse. I believe in the lectures here—or at least in the book—[you] calculated the orbit by numerical methods and saw that it looked like an ellipse. That's not exactly the same thing as *proving* that it is exactly an ellipse. The Mathematics Department ordinarily is left the job of proving that it's an ellipse, so that they have something to do over there with their differential equations. [Laughter]

I prefer to give you a demonstration that it's an ellipse in a completely strange, unique, [and] different way than you are used to. I am going to give what I will call an elementary demonstration. [But] "elementary" does not mean easy to understand. "Elementary" means that very little is required to know ahead of time in order to understand it, except to have an infinite amount of intelligence. It is not necessary to have knowledge but to have intelligence, in order to understand an elementary demonstration. There may be a large number of steps that are very hard to follow, but each step does not require already knowing calculus, already knowing Fourier transforms, and so on. So by an elementary demonstration I mean one that goes back as far as one can with regard to how much has to be learned.

Of course, an elementary demonstration in this sense could be first to teach [you] calculus and then to make the demonstration. This, however, is longer than a demonstration which I wish to present. Secondly, this demonstration is interesting for another reason—it uses completely geometrical methods. Perhaps some of you were delighted in geometry in school with the fun of trying or having the ingenuity to discover the right construction lines. The elegance and beauty of geometrical demonstration is often appreciated by lots of people. On the other hand, after Descartes, all geometry can be reduced to algebra, and today all mechanics and all these things are reduced to analysis with symbols on pieces of paper and not by geometrical methods.

On the other hand, in the beginning of our science—that is, in the time of Newton—the geometrical method of analysis in the historical tradition of Euclid was very much the way to do things. And as a matter of fact, Newton's *Principia* is written in a practically completely geometrical way—all the calculus things being done by making geometric diagrams. We do it now by writing analytic symbols on the blackboard, but for your entertainment and interest I want you to ride in a buggy for its elegance, instead of in a fancy automobile. So we are going to derive this fact by purely geometrical arguments—well, by essentially geometrical arguments, because I don't know what that means, anything precise I don't know what it means, like purely geometrical arguments—but essentially geometrical arguments, and see how well we get on.

So our problem is to demonstrate that if this is true—that the changes in velocities are directed toward the Sun, and they are inversely as the square of the distance in equal times—that the orbit is an ellipse. We then have first to understand—we must start with something—we first must know what an ellipse is. If there is no available definition of an ellipse, it is going to be impossible to demonstrate the theory. And furthermore, if you cannot understand the meaning of this proposition, of course you also cannot demonstrate the theorem. So, many people have said, "Oh yeah, but you've got to know something about an ellipse." I know—you can't state the statement otherwise. And also you have to have some understanding of this idea. That's also true. But beyond that, I don't think we need much extra knowledge, but a large amount of attention, please, and careful thinking. That's not easy, and it's quite a job, and it's not worthwhile. It is much easier to do it by the calculus, but you're going to do it that way anyway, and you must remember that this is just to see how it would look.

There are several ways of defining an ellipse, and I have to choose one, and I will suppose that the one with which everyone is familiar is the fact that an ellipse can be made, or the ellipse is the curve that can be made, by taking one string and two tacks and putting a pencil here and going around. Or mathematically, it is the locus (nowadays they say the set of all points)—all right, the set of all points—such that the sum of the distance *FP* and the distance *F'P* [*F* and *F'* being] the two fixed points, remains constant. I suppose you know that's the definition of an ellipse. You may have heard another definition of an ellipse: if you wish, these two points are called the foci, and this focus means that light emitted from *F* will bounce to *F'* from any point on the ellipse.

Let me just demonstrate the equivalence of those two propositions, at least. So the next step is to demonstrate that light will be reflected from *F* to *F'*. The light is reflected as though the surface here were a plane tangent to the actual curve. What I therefore have to demonstrate is this—and you know, of course, that the law of reflection for light from a plane is that the angle[s] of incidence and reflection are the same. Therefore, what I have to prove is this: that if I were to draw a line here, such that its angles made with the two lines *FP* and *F'P* are equal, that that line is then tangent to the ellipse.

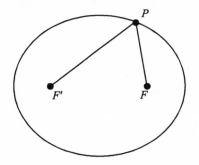
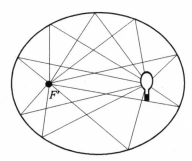

Proof: Here's the line drawn as described. Make the image point of *F'* in this line. That is to say, extend the perpendicular from *F'* to the line the same distance on the other side, to obtain *G'*, the image of *F'*. Now connect the point *P* to *G'*. Notice [that] because of the equal angles, that this angle here is the vertical angle. Well, this angle is equal to this angle, because these two right triangles are exactly the same. It's an image, so this side is the same as that side, and these two angles are equal; this is a straight line. So that *PG'* here is exactly equal to the *F'P* part, and incidentally, *FG'* is a straight line, so that the *FP* + *F'P*, which is the sum of these two distances, is in fact *FP* + *G'P*, because $F'P = G'P$. Now, the point is that if you take any other point on the tangent—say, Q—and you took the sum of these two distances to Q, it is easy to see that the distance $F'Q$ is, again, the same as $G'Q$. So that the sum of these two distances, $F'Q$ to F, is the same as the distance from F to Q and Q to G'. In other words, the sum of the distances from the two foci on any point on the line is equal to the distance from F to G', by going up to that point and across. Evidently larger, evidently always larger than going on the straight line across. In other words, the sum of the two distances to a point Q is greater than it is for the ellipse—for any point Q except for point P. For any point on this line, then, the sum of the distances to these two points is greater than it is for a point on the ellipse.

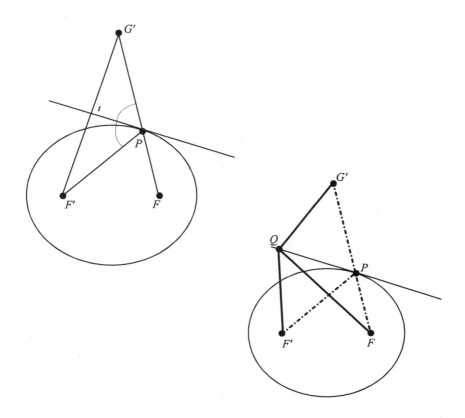

Now I take the following to be evident and perhaps you can devise a proof to satisfy you—that if the ellipse is the curve in which the sum of the two points is a constant, that the points outside the ellipse have the sum to the two points greater and the points inside the ellipse have the sum to the two points less; so that since these points on the line have a sum greater than a point on the ellipse, all this line lies outside the ellipse with the sole exception of the point P, whence it must be tangent and does not intersect at two points nor ever come inside. All right, so the thing is therefore tangent, and we know that the reflection law is right.

I have another property to describe about an ellipse, the reason for which will be completely obscure to you, but it's something which I will need later in this demonstration.

May I say that although the methods of Newton were geometrical, he was writing in a time in which the knowledge of the conic sections was the thing that everybody knew very well, and so he perpetually uses (for me) completely obscure properties of the conic sections, and I have, of course, to demonstrate my properties as I go along. I would like, however, for you to take the same diagram again, which I made here, and draw it over again. It's drawn exactly the same here: F' and F, there's that tangent line, here's the image point G' of F'. However, I would like for you to imagine what happens to the image point G' as the point P goes around the ellipse. It is evident, as I already indicated, that PG' is the same as $F'P$, so that $FP + F'P$ is a constant, [and that] means that $FP + PG'$ is a constant. In other words, that FG' is a constant. In short, the image point G' runs around the point F in a circle of constant radius. All right. At the same time, I draw a line from F' to G' and I find [that] my tangent is perpendicular to it. That's the same statement as all that was before. I just want to summarize that, to remind you of a property of an ellipse, which is this: that as a point G' goes around a circle, a line drawn from an eccentric point to this point G'—this is an off-center point to the point G'—will always be perpendicular to the tangent of the ellipse. Or the other way around: the tangent is always perpendicular to the line—or a line—drawn from an eccentric point. All right, that's all, [and] we'll come back to it and we'll remember, and we will review it again, so don't worry. That's just a sum-

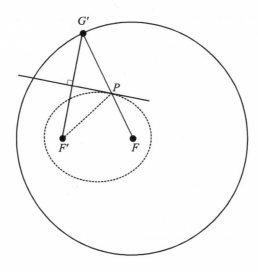

mary of some of the properties of an ellipse, starting from the facts. That's the ellipse.

On the other hand, we have to learn dynamics, we have to put them together. So now we have to explain what dynamics is all about. I want this proposition, that's the geometry; now the mechanics, what this proposition means. What Newton means by this is this: that if this is the Sun, for instance, the center of the attraction, and at a given instant a particle were to, say, be here, and let me suppose that it moves to another point, from A to B, in a certain interval of time. Then, [if] there were no forces acting toward the Sun, this particle would continue in the same direction and go exactly the same distance to a point c. But during this motion there's an impulse toward the Sun, which, for the purposes of analysis, we will imagine all the curves at the middle instant—in other words, at this instant. In other words, we concentrate all our impulses in an approximate way of thinking to this middle moment. And, therefore, the impulse is in the direction of the Sun, and this might represent the change in motion. That means that instead of this moving to here, it moves to a new point, which is C, which is different than c, because the ultimate motion is this motion compounded from the original plus the additional impulse given toward the center

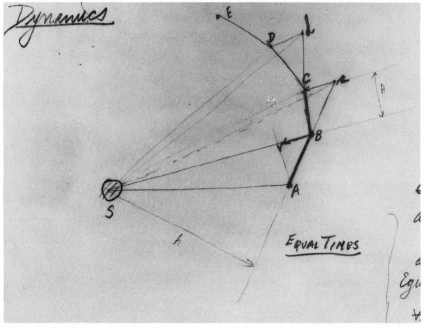

Diagram from Feynman's lecture notes.

of the Sun. So that the ultimate motion is along the line *BC*, and at the end of the second interval of moment of time the particle will be at *C*. I emphasize that *Cc* is parallel to and equal to *BV*, let us say, the impulse given from the Sun. It is therefore parallel to a line from *B* to the center of the Sun. Finally, the rest of the statement is that the size of *BV* will vary inversely as the square of the distance as we go around the orbit.

I have drawn this same thing over again here—exactly the same way, no change at all, excepting color makes it more interesting. Here's the motion that the particle would have—has in the first instant of time— and the motion which it would continue to have if it were to continue for the second interval of time with no force. May I point out to you that the areas that would be swept through in that case would be equal during those two intervals of time. For these two distances, *AB* and *Bc*, are evidently equal, and therefore the two triangles *SAB* and *SBc*, which are the two areas, will be equal: for they have equal bases and a common altitude. If you extend the base and draw the altitude, it's the same altitude for both triangles; and since the bases are equal, the areas then swept through are equal.

On the other hand, the *actual* motion is not to the point c but to the point C, which differs from the position c by a displacement in the direction of the Sun at the moment B, that is, in the blue line parallel to the original blue line. Now I would like to point out to you that the area that would be most occupied—I mean, which would be swept out in that second interval of time even if there were a force: namely, the area SBC—is the same as the area that there would be if there were no force—namely, SBc. The reason is that we have two triangles which

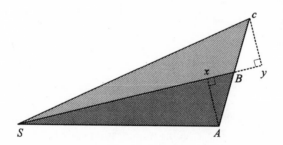

Figure in Chapter 3

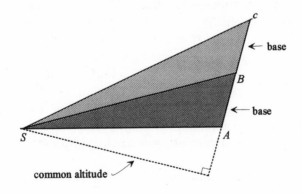

Feynman does it this way instead

have a common base and who have an equal altitude, for they lie between parallel lines. Since the area[s] of the triangle *SBC* and the triangle *SAB* are equal—but since those points *A*, *B*, and *C* represented positions in succession at equal times in the orbit—we see that the area[s]moved through in equal times are equal. We can also see that the orbit remains a plane, that the point *c* being in the plane and the line *Cc* being in the plane of *ABS*, the remaining motion is in the plane *ABS*.

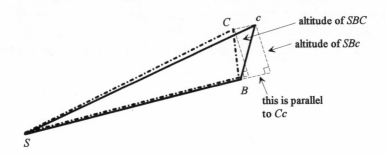

And I have drawn a succession of such impulses around this imaginary polygonal orbit. Of course, to find the actual orbit, we need to make the same analysis with a much smaller interval of time—and a much finer rate of impulsing—until we get the limiting case, in which we have a curve. And in the limiting case in which we have a curve—the area swept by this thing—the curve will lie in a plane, and the area swept will be proportional to the time. So that's how we know that we have equal areas in equal times. The demonstration that you have just seen is an exact copy of one in the *Principia Mathematica* by Newton, and the ingenuity and delight which you may or may not have gotten from it is that already existing in the beginning of time.

Now the remaining demonstration is not one which comes from Newton, because I found I couldn't follow it myself very well, because it involves so many properties of conic sections. So I cooked up another one.

We have equal areas and equal times. I would like now to consider what the orbit would look like if instead of using equal time, one were to think of the succession of positions which correspond to *equal angles* from the center of the Sun. In other words, I repicture the orbit with the succession of points, *J*, *K*, *L*, *M*, *N*, which correspond not to equal instants, like they did in the diagram before, but rather [to] equal angles of inclination from the original position. To make this a little bit simpler, although it is not at all essential, I have supposed that the original motion was perpendicular to the Sun at the first point—but that's not essential, it just makes the diagrams cleaner.

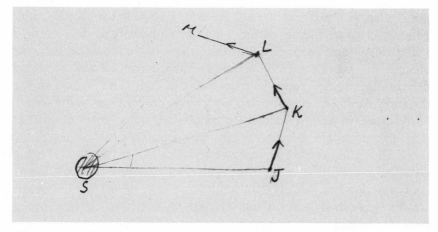

Diagram from Feynman's lecture notes.

Now we know from the proposition previously that equal [areas] occupy equal times to be swept through. Now listen: I would point out to you that . . . equal angles, which is what I'm aiming for, means that areas are not equal, no, but they are proportional to the square of the distance from the Sun; for if I have a triangle of a given angle, it is clear that if I make two of them that they are similar; and the proportional area of similar triangles is proportional to the square of their dimensions.[2] Equal angles therefore means—since areas are proportional to time—equal angles therefore means that the times to be swept through these equal angles are proportional to the square of the distance. In other words, these points—*J*, *K*, *L*, and so on—do not represent

[2]This is the point explained in the footnote to Chapter 3, page 115.

pictures of the orbit at equal times, no, but they represent pictures of the orbit with successions of times which are proportional to the square of the distance.

Now, the dynamical law is that there are equal changes in velocity, no—that the changes in velocity vary inversely as the square of the distance from the Sun—that is, the changes of velocity in equal times. Another way of saying the same thing is that equal changes of velocity will occupy times proportional to the square of the distance. It's the same thing. If I take more time, I get more change in the velocity, and, although they are falling off for equal times inversely as the square, if I make my times proportional to the square of the distance, then the changes in velocity will be equal. Or, the dynamical law is: equal changes in velocity occur in times proportional to the square of the distance. But look, equal angles were times proportional to the square of the distance. And so we have the conclusion, from the law of gravitation, that equal changes of velocity will occur in equal angles in the orbit. That's the central core from which all will be deduced—that equal changes in velocity occur when the orbit is moving through equal angles. So I now draw on this diagram a little line to represent the velocities. Unlike the other diagram, those lines are not the complete line from J to K, for in that diagram those were proportional to the velocities, for the times were equal, and the length divided by equal times represented the velocities. But here I must use some other scale to represent how far the particle would have gone in a given unit of time, rather than in the times which are, in fact, proportional to the square of the distance. So these represent the velocities in succession. It is quite difficult in that diagram to find out what the changes are.

I therefore make another diagram over here, which I'll call the diagram of the velocities, in which I draw a picture on a magnified scale only for convenience. These are supposed to represent exactly these same lines. This would represent the motion per second of a particle at J or in a given interval of time, at J. This would represent the motion that a particle would've made from the beginning in a given interval of time. And, I put them all at a common origin, so that I can compare the velocities. So I have then a series of the velocities for the succession of these points.

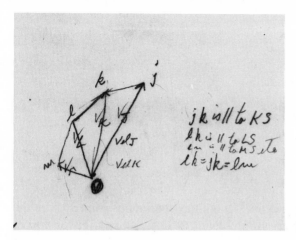

Diagram from
Feynman's lecture notes.

Now, what are the changes in the velocity? The point is that in the first motion, this is the velocity. However, there is an impulse toward the Sun, and so there is a change in velocity, indicated by the green line that produces the second velocity, v_K. Likewise, there's another impulse toward the Sun again, but this time the Sun is at a different angle, which produces the next change in the velocity, v_L, and so on. Now, the proposition that the changes in the velocities were equal—for equal angles, which is the one that we deduced—means that the lengths of these succession of segments are all the same. That's what it means.

And what about their mutual angles? Since this is in the direction of the Sun at this radius, since this is at the direction of the Sun at that radius, and since this is the direction of the Sun at that radius, and so on, and since these radii each successively have a common angle to one another—so it is likewise true that these little changes in the velocity have, mutually to one another, equal angles. In short, we are constructing a regular polygon. A succession of equal steps, each turn through an equal angle, will produce a series of points on the surface underlying a circle. It will produce a circle. Therefore, the end of the velocity vector—if they call it that, the ends of these velocity points; you're not supposed to know what a vector is in this elementary description—will lie on a circle. I draw the circle again.

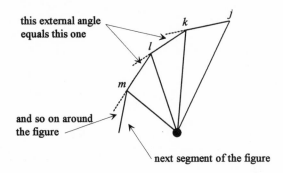

this external angle
equals this one

k   *j*

l

m

and so on around
the figure

next segment of the figure

I review what we found out. I take the continuous limit, where the
intervals of angle are very tiny indeed, to obtain a continuous curve.
Let θ be the angle, total angle, to some point P, and let v_P represent
the velocity of that point in the same way as before. Then the diagram
of velocities will look like this. This is the origin of the velocity diagram,
the same as over there, and this is the velocity vector corresponding to
this point P. Then this lies on a circle, but always not necessarily the
center of that circle. However, the angle that you've turned through in
the circle is the same θ as here. The reason for that is that the angle
turned through from the beginning by this thing is proportional to the
angle turned through by the orbit, because it's the succession of the
same number of small angles. And therefore, this angle in, here, is the
same angle as in, here.

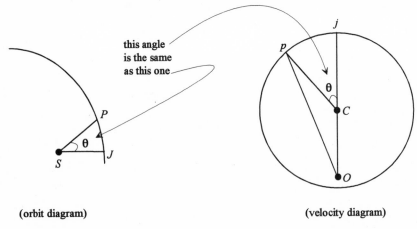

this angle
is the same
as this one

P

θ

S   *J*

j

p

θ

C

O

(orbit diagram)   (velocity diagram)

So here is the problem, here's what we have discovered: that if we draw a circle and take an off-center point, then take an angle in the orbit—any angle you want in the orbit—and draw the corresponding angle inside this constructed circle and draw a line from the eccentric point, then this line will be the direction of the tangent. Because the velocity is evidently the direction of motion at the moment and is in the direction of the tangent to the curve. So our problem is to find the curve such that if we draw a point from an eccentric center, the direction of the tangent of that curve will always be parallel to that when the angle of the curve is given by the angle in the center of that circle.

In order to make still clearer why it is going to come out in this thing, I'll turn the velocity diagram 90°, so that the angles correspond exactly and are parallel to each other. This diagram under here, then, is precisely the same diagram as the one you see above, but turned 90°—only to make it easier to think. This, then, is the velocity vector, except that it's turned 90° because the whole diagram is turned 90°. That is, this is perpendicular evidently to that, and therefore this is evidently perpendicular to that. In short, we must find the curve such that if we put the orbit in it, I think I've started—yes, so I'll just say it and then I'll draw it again—if we put the orbit in it at a given point, here, where this line intersects the orbit (never mind the scales, they're all imaginary, I mean, it's all in proportion), where this line intersects the orbit, the tangent should be perpendicular to that line from an eccentric point.

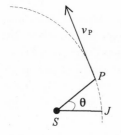

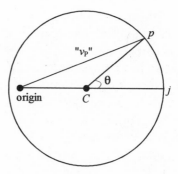

I draw it again, to show you how it is. You know now what the answer is. But here's a picture again of the same velocity circle, but this time the orbit is drawn inside at a different scale, so that we can see this picture laid right over this picture, so the angles correspond. So since the angles correspond, I can draw the single line to represent both the point P on the orbit and the point p on the velocity circle. Now what we have discovered is that the orbit is of such a character that a line drawn from the eccentric point—here, from an extension of this point onto a circle outside—will always be perpendicular to the tangent to the curve. Now that curve is an ellipse, and you can find that out by the following construction.

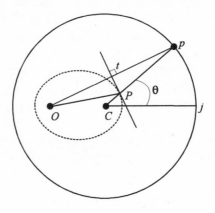

Construct the following curve. The curve I'm going to construct will satisfy all the conditions. Construct the following curve. Always take the perpendicular bisector of this line and ask for its intersection with the other line, Cp, and call that intersection point P. This is the perpendicular bisector. Now I'll prove two things. First, that the locus of this point that's been generated there is an ellipse, and, second, that this line is a tangent thereto—that is, to the ellipse—and therefore satisfies the conditions, and all is well.

First, that it's an ellipse: Since this was the perpendicular bisector, it is at equal distances from O and p. It is therefore clear that Pp is equal to PO. That means that CP + PO, which is therefore equal to

$CP + Pp$, is the radius of the circle, which is evidently constant. So the curve is an ellipse, or the sum of these two distances is a constant.

And next, this line is tangent to the ellipse because, since . . . the two triangles are congruent, this angle here is equal to this angle here.

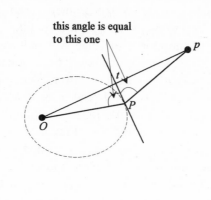

this angle is equal
to this one

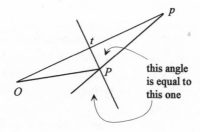

this angle
is equal to
this one

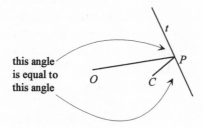

this angle
is equal to
this angle

But if I extend this line on the other side, [then] also is that angle equal. So therefore the line in question makes an equal angle with the two lines to the foci. But we proved that that was one of the properties of an ellipse—the reflection property. Therefore, the solution to the problem is an ellipse—or the other way around, really, is what I proved: that the ellipse is a possible solution to the problem. And it is this solution. So the orbits are ellipses. Elementary, but difficult.

I have considerable more time, and so I will say a few things about this. In the first place, I would like to say how I got this demonstration—the fact that the velocities went in a circle. The demonstration [of] this point was due to Mr. Fano and I read it. And after that, to prove that it was an ellipse took me an awful long time: that is, the obvious, simple step—you turn it this way, and you draw that and all that. Very hard, and like all these elementary demonstrations they require a large amount—like any geometrical demonstration—of ingenuity. But once presented, it's elegantly simple. I mean, it's just finished. But the fun of it is that you've made a kind of a carefully put-together piece of pieces.

It is not easy to use the geometrical method to discover things. It is very difficult, but the elegance of the demonstrations after the discoveries are made is really very great. The power of the analytic method is that it is much easier to discover things than to prove things. But not in any degree of elegance. It's a lot of dirty paper, with x's and y's and crossed out, cancellations and so on.

I would like to point out a number of interesting cases. It of course can happen that the point O lies on the circle, or even that the point O lies outside the circle. It turns out that the point O lying on the circle does not produce, of course, an ellipse; it produces a parabola. And the point O lying outside the circle, which is another possibility, produces a different curve, a hyperbola. I leave some of those things for you to play with. On the other hand, I would like now to make some application of this and to continue the argument that Mr. Fano originally made, for another purpose. He was going in a different direction, and I'd like to show you that.

What he [Fano] was trying to do was to make an elementary demonstration of a law which was very important in the history of physics in

1914. And that had to do with the so-called Rutherford's law of scattering. If we have an infinitely heavy nucleus—which we don't have, but suppose—and if we shoot a particle by that nucleus, then it will be repelled by an inverse-square law, because of the electrical force. If q_e is the charge on an electron, then the charge on the nucleus is Z times q_e when Z is the atomic number. Then the force between the two things is given by $4\pi\epsilon_0$ times the square of the distance, which for simplicity I will write temporarily as z/R^2—the constant over R^2. I don't know whether you've done this in the class or not; but I'll suppose, I'll define another thing because, $q_e^2 / 4\pi\epsilon_0$ will be written e^2 for short. Then this thing is just Ze^2 / R^2. Anyway, that's the force inversely as the square of the distance, but it's a repulsion. And now the problem is the following: If I shoot a lot of particles at these nuclei, where I can't see the nuclei, how many of them will be deflected through various angles? What percentage will be deflected more than 30°? What percentage will be deflected more than 45°? And how are they distributed in angles? And that was the problem that Rutherford wanted solved, and when he had the correct solution, he then checked it against experiment.

[At this point, Feynman goes off in the wrong direction. He'll correct himself in a moment.]

And he found that the ones that were supposed to be deflected through large angles were not there. In other words, the number of particles deflected through large angles was much less than you would think, and he therefore deduced that the force was not as strong as $1/R^2$ for small distances. Because it is obvious that to get the large angle, you need a lot of force, and it corresponds to the [particles] that hit [the nucleus] almost head-on. So those which come very close to the nucleus do not seem to come out the way they ought to, and the reason is that the nucleus has a size . . . I've got the story backwards. If the nucleus had a big size, then those which were supposed to come out at large angles wouldn't get their full force, because they would get inside the charge distribution and would be deflected less. I got mixed up. Excuse me. I start again.

Rutherford deduced how it should go if all the forces were concentrated at the center. In his day, it was supposed that the charge in an atom was distributed uniformly over the atom, and in order to discover this distribution, he thought that if he scattered these particles, they would show a weaker deflection—they would never show a very large deflection corresponding to a very close approach to the repulsion center because [in] the close approach there's no center. He, however, did find the large-angle deflections, and deduced that the nucleus was small and that the atom had all its mass at a very small central point. I got it backwards. It was later that it was demonstrated, by the same thing again, that the nucleus has a size. But the first demonstration was that the atom is not as big, for this kind of electrical purposes, as the whole atom is known to be: that is, all the charge was concentrated at the center, and thus the nucleus was discovered. However, we need now to understand this: we need to know what the law is for the angle of deflection here, and that we can obtain in this way.

Suppose that we do the same thing as we did before, and we draw the orbit. Here is the charge, and here is the motion of a particle going around, only this time it's repulsion. I start the picture at this point, for the fun of it, and I draw my velocity circles as before. This is the velocity. We know that the velocity, the initial velocity at this point—I should use the same colors so you know what I'm doing, this should be blue, this orbit is red—now the velocity changes lie on a circle. But the changes in the velocity this time are repulsions, and the sign is reversed. And after some minor thought, you can see that the deflections go like that, and that the center of the calculation [which] used to be called the origin of the velocity space O, lies on the outside of the circle. And the succession of small velocity changes lie on the circle, and the succession of velocities then in the orbit are these lines, until a very interesting point comes: until we get to this tangent.

At this tangent point to the curve—what does it mean? It means that all the changes in velocity are in the direction of the velocity. But the changes in the velocity are in the direction of the Sun, and that means that this velocity, in this part of the diagram, is in the direction of the Sun, because it is in the direction of the changes. That is to say, this point here, as we approach this point here—which I could call x, say—corresponds to coming from infinity toward the Sun along a line here. That is, very far out we are directed toward the Sun very closely (not the Sun, but the nucleus) and then as it comes around here—this diagram

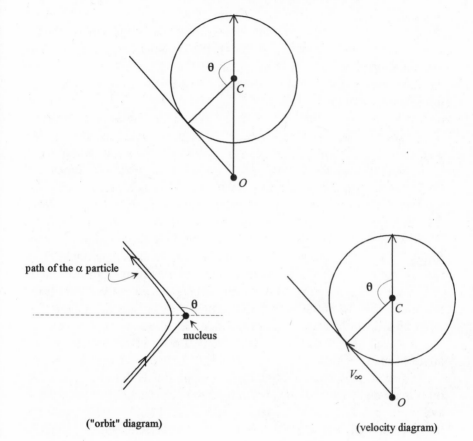

("orbit" diagram)   (velocity diagram)

should be the other way, the arrows should be here, I got the changes the wrong way in time—comes around here and goes out this way and, going out that way, corresponds to going with the velocity off in this direction.

Now, if we draw then the orbit more carefully, it will look very much like this. It goes around like this. If I call this point, here, V_∞ then the velocity that the particle has at the beginning is V_∞. If, on the same scale, I call the radius of this circle V—the velocity corresponding to the radius of the circle—I'm going to make up some equation, I'm not going to do it completely geometrically, but to save time and so on, I've done all the work. One should not ride in the buggy all the time. One has the fun of it and then gets out. Now first I want to find the velocity of the center, the radius of the velocity circle. In other words, I'm now going to come down and make some of these geometric things more analytic.

I will suppose that the force is some constant: the force—the acceleration, rather—is some constant over R^2. For gravity, this constant is GM and, for electricity, it is Ze^2/m, over m because of the acceleration. That is to say, the changes in velocity are always equal to z/R^2 times the time. Now let us suppose that we call α, which is a constant for the motion, the area swept by the orbit per second. That is then this way: that the time—if I wanted to change this to angle, I have the following— $R^2\Delta\theta$ would be the area. If I divide that by the rate that area is swept through—this tells me how much time it takes to sweep an angle. The time is, then, for given angles, proportional to the square of the distance. All this I'm saying now analytically, where I said in words before. Substitute this Δt, in here, to find out how the changes in the velocity are with respect to angle, and one obtains $R^2\Delta\theta/\alpha$, or the R^2's cancel, and it means that the changes in velocity are as advertised: for equal angles, equal.

Now then, the velocity diagram—although this isn't the piece of the orbit that you can get to, never mind—these are changes in the velocity and these are changes in the angle in the orbit. So ΔV is also equal by the geometry of that circle to the radius of the circle, which I call V_R $\times$ $\Delta\theta$. In other words, we have that the radius of the velocity circle is equal to z/α, where α is the rate of area swept per second and z is a constant having to do with the law of force. Now, the angle through which this planet has deflected is this one, here, and I call it, the angle of deflection from the planet—I mean the charged particle from the nucleus. It is evident from my discussion that it's the same as this angle in here, ϕ, because these velocities are parallel to the two original directions. It is clear, therefore, that we can find ϕ if we can get the relation with V_∞ and V_R. You see, look, tangent of $\phi/2 = V_R/V_\infty$ and that gives us the angle. The only thing is that we need—we have to substitute for V_R, $z/\alpha R$, and we have that much.

Now, it doesn't do us much good until we know α for this orbit. An interesting idea is this: think of this thing as approaching this, so that if there were no force it would miss by a certain distance, b. This is called the impact parameter. We imagine that the thing comes from infinity aimed for the force center, but is missing—because it misses, it is deflected. By how much is it deflected, if it was aimed to miss by b? That's the question. If it's aimed to miss by a distance b, how much will it get deflected?

So I need now only determine how α is related to b. V_∞ is the distance gone in 1 second, so if I were to draw way out here a horrible-looking area, a triangle—a terrible-looking triangle, then the—I got a factor of 2 somewhere, yeah, the area of a triangle is $1/2$ R^2. There are two factors, two, which you will straighten out please when the time comes. There is $1/2$ in here and, there is $1/2$ somewhere else, which I'm now going to make. The area of this triangle is the base V_∞ times the height b times $1/2$. Now that triangle is a triangle through which a particle would sweep—the radius would sweep in 1 second. And this is, therefore, α. So, therefore, we have that this goes as z/bV_∞^2. That tells us that given the impact distance, the aiming accuracy, what angle we would find in the deflection in terms of the speed at which the particle approaches and the known law of force. So it's completely finished.

One more thing that is rather interesting. Suppose that you would like to know with what probability, what chance is there of getting a deflection more than a certain amount. Let's say you pick a certain ϕ— ϕ_0, say—and you want to make sure that you get greater than ϕ_0. That only means that you have to hit inside an area closer than the b which belongs to that ϕ. Any collision closer than b will produce a deflection bigger than ϕ_0, where b is b_0, belonging to ϕ_0 through this equation. If you come further away, I have less deflection, less force. So, therefore, the so-called cross section of area that you have to hit for deflection, to be greater than ϕ (I'll leave off the naught), is πb^2, where b is z/V_∞^2 $\tan^2 \phi/2$. In other words, it is $\pi z^2/V_\infty^4 \tan^2 \phi/2$. And that's the law of Rutherford's scattering. That tells you the probability of the area you have to hit—the effective area that you have to hit—in order to get a deflection more than a certain amount. This z is equal to Ze^2/m; this is a fourth power, and it is a very famous formula.

It is so famous that, as usual, it was not written in this form when it was first deduced, and so I, just for the famousness of it, will write it in a form—well, I'll leave you to write it in a form. I'll write just the answer, and I'll let you see if you can show it. Instead of asking for the cross section for a deflection greater than a certain angle, we can ask for the piece of cross section, $d\sigma$, that corresponds to the deflection in the range $d\phi$ that the angle should be between, here, and there. You just have to differentiate this thing, and the final result for that thing is given as the famous formula of Rutherford, which is $4\,Z^2e^4$ times $2\pi\sin\phi\,d\phi$ divided by $4m^2\,V_\infty^4$ times the sine of the fourth power of $\phi/2$. This I write only because it's a famous one that comes up very much in physics. The combination $2\pi\sin\phi\,d\phi$ is really the solid angle that you have in range $d\phi$. So in a unit of solid angle, the cross section goes inversely as the fourth power of the sine of $\phi/2$. And it was this law which was discovered to be true for scattering of α particles from atoms, which showed that the atoms had a hard center in the middle . . . a nucleus. And it was by this formula that the nucleus was discovered.

Thank you very much.

Epilogue

Richard Feynman conjured up his own brilliant proof of the law of ellipses, but he was not the first to think of it. The same proof, right down to the crucial insight of turning the velocity diagram on its side, appears in a little book called *Matter and Motion*, written by James Clerk Maxwell and first published in 1877. Maxwell attributes the method of proof to Sir William Hamilton, a name familiar to all physicists. (The Hamiltonian is a crucial element of quantum mechanics.) Apparently, Hamilton was the first to use the velocity diagram, which he called the Hodograph, to study the motion of a body. In his lecture, Feynman generously credits a mysterious "Mr. Fano" with the idea of the circular velocity diagram. He is referring to a book by U. Fano and L. Fano, *Basic Physics of Atoms and Molecules* (1959), where a circular velocity diagram is used to derive the Rutherford scattering law presented by Feynman at the end of his lecture. If Fano and Fano knew about Hamilton and his Hodograph, they do not say so.

Hamilton was part of a centuries-long tradition of refining Newton's mechanics into formulations of ever greater sophistication and elegance. For more than two hundred years after the publication of the *Principia*, the universe of Newton reigned supreme. Then, early in the twentieth

171

century, a second scientific revolution took place in physics, almost as far-reaching as the first one. When it was over, Newton's laws could no longer be regarded as revealing the innermost nature of physical reality.

The second revolution took place on two separate fronts that have not yet, even today, been fully reconciled. One led to the theory of relativity. The other led to quantum mechanics.

The seeds of the theory of relativity can be traced as far back as Galileo's discovery that all bodies fall at the same rate regardless of mass. Newton's explanation was that the mass of a body plays two separate roles in physics: one role is to resist changes in the motion of the body; the other is to apply gravitational force to the body. Thus, the greater the mass of a body, the stronger the force of gravity on it, but also the more difficult to get it moving. Heavier bodies—falling toward the Earth, for example—have greater force on them but more strongly resist being accelerated. Lighter bodies have smaller forces but are more easily accelerated. The net effect is that all bodies fall at exactly the same rate. This peculiar coincidence was easy to accept as part of the price for the vast success of Newton's mechanics.

By the end of the nineteenth century, however, another part of Newton's laws had come into question as a result of the discoveries of none other than James Clerk Maxwell. It had long been known that light is not transmitted instantaneously but rather travels at a definite speed. That speed is very great—roughly 186,000 miles (or 300,000 kilometers) per second—but it is not infinite. It was also known by Maxwell's era (he lived from 1831 to 1879, the year of Einstein's birth, and died, like Feynman, of stomach cancer) that while electricity is a force acting between electric charges, magnetism, the force that orients compass needles, is not a completely separate phenomenon. Instead, magnetism is a force between electric currents, and electric currents are simply electric charges that are moving. Maxwell discovered that if you compared the strength of the electric force between charges at rest with the strength of the magnetic force between slowly moving charges, the ratio was equal to the square of a velocity that just happened to be the same as the speed of light! Maxwell knew that this was no mere coincidence, and he worked out an elegant mathematical theory, quickly confirmed by experiment, that all space is permeated by electric and magnetic fields of force, and that when these fields are disturbed, the disturbance propagates at the speed of light; in fact, the disturbance *is* light.

It was not immediately obvious that this discovery undermined Newton's laws, but it was soon realized, by Albert Einstein, that it did just that. In the old Aristotelian world, the natural state of a body is rest. In Newton's world, there is no such thing as a state of absolute rest. A body tends to stay in motion, at constant speed in a straight line. If a body seems to be at rest, that is only because the observer is moving together with it. Newton's first law, the law of inertia, makes sense because there is no such thing as a state of rest. In a universe in which there is no state of rest—where one state of motion is as good as any other—the simplest assumption possible is that a body will retain whatever state of motion it has, which is precisely what the law of inertia says. However, if there is no absolute rest, there should be no absolute speed. The apparent speed of anything should depend on whether the observer is moving along with it or not. That's where the crunch comes: the laws of physics should never have in them a definite speed, because the speed of anything should depend on the speed of the observer. But James Clerk Maxwell had shown that light has a definite speed—a speed that can be found in the fundamental forces between magnets and between electric charges.

To resolve this anomaly, Albert Einstein created a whole new universe. Its central axioms, from which all else is deduced, are that there is a single absolute speed of light, regardless of the speed of the observer, and that all bodies, regardless of their mass, fall at the same rate because the pull of gravity downward on a body is indistinguishable from an upward acceleration of everything but the body. To assure that the speed of light is the same for all observers, time and distance must lose their independent, Newtonian meanings and mix together into spacetime. To make all bodies fall at the same rate, the very force of gravity itself is replaced by curved spacetime, in which all bodies move inertially—not on straight lines (no such thing exists anymore) but along curves called geodesics, which are the shortest distance between two points in the curved spacetime. All of this is known collectively as the theory of relativity (both special and general).

The other front of advancing knowledge that undermined the supremacy of Newton was the nature of the atom. The existence of atoms had been suspected at least since the time of Lucretius, in the first century B.C., believed in by most scientists, including Newton, and finally given some empirical support at the dawn of the nineteenth century by the English chemist John Dalton. Dalton did experiments in which he claimed to show that chemical species, such as nitrogen and oxygen, tend to combine in ratios of simple whole numbers (one-to-one, one-to-two, two-to-three, and so on; the quantities were measured by volume in the gaseous state). These experimental results clearly implied that the constituents of the gases were atoms, combining into what we would today call simple molecules (NO, NO_2, N_2O_3, and so on). Dalton, who was an inept experimenter but a firm believer in atoms, announced his discovery on the basis of very poor evidence (a story not uncommon in the history of science), but more skillful chemists went on to make his law of simple and multiple proportions one of the central doctrines of experimental chemistry. Throughout the nineteenth century, knowledge of the properties of atoms was gradually refined. The 1875 edition of the *Encyclopaedia Britannica* has, under the heading "Atoms," a superb review of the state of knowledge at the time, signed "JCM"— for James Clerk Maxwell. The next real breakthrough, however, came in 1896, when the English physicist J. J. Thomson was able to show that all atoms share a common internal constituent that came to be called the electron.

At this point, the issue became the architecture of the atom. The experiment by Ernest Rutherford and his colleagues, described in Feynman's lecture, implied that the atom was a kind of miniature solar system, with a tiny but heavy nucleus at the center and lightweight electrons in orbit around it, which were held in place not by gravity but by the electric force between their own negative charge and the positively charged nucleus. However, this comforting view of a tiny Newtonian solar system in every atom had a number of fundamental flaws, chief among them an absolute prohibition due, once again, to James Clerk Maxwell and his theory of electromagnetism. If electrons were indeed in orbit around the nucleus, they would continually disturb the electromagnetic field. That disturbance would propagate away at the speed of light, draining energy out of the atom until it collapsed, the electrons expiring by falling into the nucleus, like tired comets falling into the Sun. Since common experience tells us that most atoms are stable and

James Clerk Maxwell

Ernest Rutherford

long-lived, the Newtonian solar system will not do as a description of the inner workings of the atom.

The solution to this dilemma was the invention of quantum mechanics. Newton's laws do not describe the behavior of the very small. As the character Kerner (a Feynman-like physicist turned spy) says in Tom Stoppard's play *Hapgood*:

> There is *no such thing* as an electron with a definite position and a definite momentum; you fix one, you lose the other, and it's all done without tricks. . . . When things get very small they get truly crazy. . . . So now make a fist, and if your fist is as big as the nucleus of one atom then the atom is as big as St. Paul's, and if it happens to be a hydrogen atom then it has a single electron flitting about like a moth in the empty cathedral, now by the dome, now by the altar. . . . Every atom is a cathedral. . . . An electron does not go round like a planet, it is like a moth which was there a moment ago, it gains or loses a quantum of energy and it jumps, and at the moment of quantum jump it is like *two* moths, one to be here and one to stop being there; an electron is like twins, each one unique, a unique twin.

Thus, early in the twentieth century, Newton was overthrown in favor of relativity and quantum mechanics, just as a couple of centuries earlier he had displaced Aristotle at the center of the intellectual universe. Why, then, do we continue to teach Newtonian physics in school? More to the point, why did Richard Feynman—the same Richard Feynman who virtually reinvented quantum mechanics, and who lectured often and brilliantly on Einstein's theory of relativity—bother to reinvent the proof of the law of ellipses by the outmoded Isaac Newton?

The answer is that the second revolution in physics was profoundly different from the first one. The first revolution overthrew Aristotelian doctrine and replaced it with something entirely different. The second revolution did not overthrow Newtonian physics in the sense of showing that it was wrong; instead, it affirmed Newtonian physics by showing why it was right. Newton's laws are no longer believed to unmask the innermost nature of physical reality; moreover, they are not even correct, if applied to things that are very small (electrons) or very fast (near the speed of light) or very dense (black holes). There are even less extreme conditions wherein departures from the predictions of Newton's laws can be detected, if we know exactly where to look. Nevertheless, the world after the second revolution is for the most part pretty much the same as the world we inhabited before it. The main difference is that we now know not only that Newton's laws give us an accurate account of how the world behaves but also *why* his laws work so well. They work because they arise naturally out of even more profound laws called relativity and quantum mechanics. Those more profound laws are needed to tell the whole story (in reality, we don't yet know the whole story), but for the most part Newton's laws do just fine.

That's the reason we still teach students how to solve problems using Newtonian physics but not Aristotelian physics. It is also the reason that Richard Feynman thought it worth his while to create his own geometric proof that Newton's laws produce elliptical orbits for the planets around the Sun. And that, finally, is the reason for this book.

Feynman's Lecture Notes

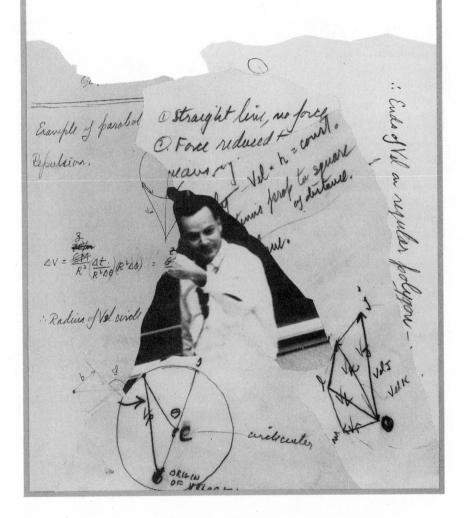

Feynman's notes for his introductory remarks, 1964.

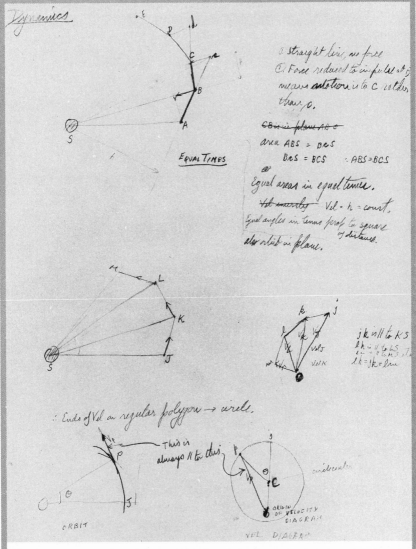

Most of the lecture comes from this page. The figure in the upper left-hand corner is copied from Newton's *Principia*.

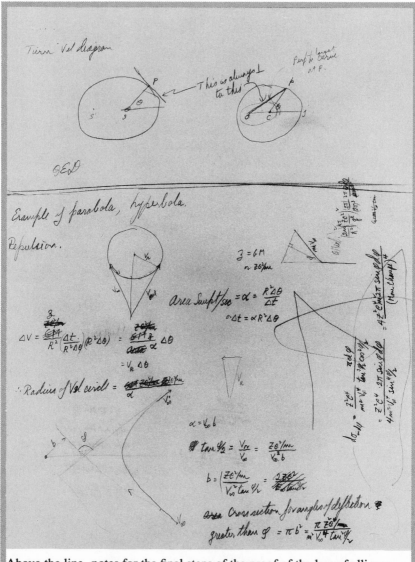

Above the line, notes for the final steps of the proof of the law of ellipses. Below the line, Rutherford's law of scattering.

Bibliography

Brecht, Bertolt. *The Life of Galileo*. Translated by Desmond I. Vesey. London: Methuen, 1960.

Cohen, I. Bernard. *The Birth of a New Physics*. Revised edition. New York: W. W. Norton, 1985.

———. Introduction to Newton's "Principia." Cambridge, England: Cambridge University Press, 1971.

Dijksterhuis, E. J. *The Mechanization of the World Picture* (1961). Translated by C. Dikshoorn. Paperback reprint, London: Oxford University Press, 1969.

Drake, Stillman. *Galileo at Work: His Scientific Biography*. Chicago: University of Chicago Press, 1978.

Fano, U., and L. Fano. "Relation between Deflection and Impact Parameter in Rutherford Scattering." Appendix III in *Basic Physics of Atoms and Molecules*. New York: John Wiley, 1959.

Feynman, R. P., R. B. Leighton, and M. Sands. *The Feynman Lectures on Physics*. 3 vols. Reading, Penn.: Addison-Wesley, 1963–65.

Galilei, Galileo. *Two New Sciences*. Translated, with introduction and notes, by Stillman Drake. Madison: University of Wisconsin Press, 1974.

———. *Il Saggiatore*. Rome: Giacomo Masardi, 1623.

———. *Dialogue Concerning the Two Chief World Systems—Ptolemaic & Copernican*. Translated by Stillman Drake. Berkeley: University of California Press, 1962.

Gingerich, Owen. *The Great Copernicus Chase and Other Adventures in Astronomical History*. Cambridge: Sky Publishing, 1992.

Kepler, Johannes. *New Astronomy*. Translated and edited by William H. Donahue. Cambridge, England: Cambridge University Press, 1992.

Koestler, Arthur. *The Sleepwalkers* (1959). Paperback reprint, New York: Grosset and Dunlap, 1963.

Maxwell, J. Clerk. *Matter and Motion* (1877). Reprint, with notes and appendices by Sir Joseph Larmor, London: Society for Promoting Christian Knowledge, 1920.

Newton, Isaac. *Sir Isaac Newton's Mathematical Principles of Natural Philosophy and His System of the World*. Edited by Florian Cajori. Berkeley: University of California Press, 1934.

Santillana, Giorgio De. *The Crime of Galileo*. Chicago: University of Chicago Press, 1955.

Stoppard, Tom. *Hapgood* (1988). Reprint, with corrections, London: Faber and Faber, 1994.

——. *Arcadia*. London: Faber and Faber, 1993.

——. "Playing with Science." *Engineering & Science* 58 (1994):3–13.

Thoren, Victor E., with John R. Christianson. *The Lord of Uraniborg: A Biography of Tycho Brahe*. Cambridge, England: Cambridge University Press, 1990.

Westfall, Richard S. *Never at Rest: A Biography of Isaac Newton*. Cambridge, England: Cambridge University Press, 1980.

Index

Page numbers in *italics* refer to photographs. Those in **boldface** refer to the text of the lost lecture.

185